AI

ETHNIC VARIABLES IN HUMAN FACTORS ENGINEERING

Ethnic Variables in Human Factors Engineering

Based on papers presented at a symposium on "National and Cultural Variables in Human Factors Engineering," held in Oosterbeek, The Netherlands, 19–23 June 1972, under the auspices of the Advisory Group on Human Factors, North Atlantic Treaty Organization

EDITED BY
ALPHONSE CHAPANIS

The Johns Hopkins University Press
Baltimore and London

Manufactured in the United States of America.

The Johns Hopkins University Press, Baltimore, Maryland 21218
The Johns Hopkins University Press Ltd., London

Library of Congress Catalog Card Number 74-24393
ISBN 0-8018-1668-8

Library of Congress Cataloging in Publication data will be found on the last printed page of this book.

Contents

Preface

During the quarter century that I have been teaching and lecturing about human factors engineering, I have frequently been asked: How general are these principles? Would these rules hold in all countries and in all cultures? Years ago, my standard reply was that science and technology are universal and that they know no geographic boundaries. Over the years, however, a considerable amount of foreign travel and practical consulting experience have brought me greater sophistication and a realization that my initial responses to these questions were far too naive.

At the same time as I began to appreciate more fully the complexities of achieving a truly international technology, it became increasingly clear that not very much was known, or had been written, about human factors from a cross-cultural standpoint. From time to time, I thought about arranging an international symposium on this topic, but most granting agencies found the subject too applied to support. In 1969, however, an old friend, Richard Trumbull, suggested that I approach the Scientific Affairs Division of the North Atlantic Treaty Organization. There my ideas at last found a sympathetic hearing. After a formal proposal and further negotiations, NATO agreed in 1971 to support a symposium on National and Cultural Variables in Human Factors Engineering.

THE SYMPOSIUM

Little did I know what I had bargained for. Finding people who had something substantive to contribute, arranging for a meeting place, overseeing all the attendant organizational problems, handling the voluminous international correspondence associated with such a project—these were only some of the administrative chores that consumed a substantial part of my time for the greater part of a year. Problems, some of an exotic nature, continued right to the end of the meetings. A couple of participants narrowly missed taking an airliner that crashed; one of our speakers was stranded for a couple of days after he accidentally threw away his airline ticket, passport, and traveler's cheques; another speaker was late because his plane had been diverted by a bomb threat.

Through all these vicissitudes, I had, fortunately, some very able assistance. Without the help of my co-organizer, Dr. John de Jong, I never could have done it. Dr. de Jong was, most important of all, in charge of local arrangements. In addition, his familiarity with the work of ergonomists the world around was an invaluable source of information. Equally important, perhaps, were his unfailing good humor and his support when I was beset with organizational trials.

On the home front, Mrs. Margaret Iwata churned out enough memoranda and letters in two languages to fill a large section of my files, mastered the intricacies of a German checking account, and kept my coffee cup filled. Overseeing and gently guiding this flurry of activity from a distant vantage point were Mr. M. A. G. Knight and his successor Dr. B. A. Bayraktar, my contacts in NATO.

In the end, the symposium was held from the 19th through the 23rd of June 1972, in Oosterbeek, The Netherlands. It was, all agreed, a huge success. Forty-four participants from fifteen countries attended. As I had defined it, "the purpose of the symposium [was] not to exchange basic information about national and cultural differences, but rather to show how and to what extent human engineering (or ergonomic) principles and practices have been, or need to be, modified to take account of national and cultural differences." It was the first symposium ever organized to deal with such a topic. The twenty-six papers presented left no doubt that national and cultural variables are, or should be, significant factors in the formulation and application of human factors principles throughout the world.

THIS BOOK

This book is, and is not, the proceedings of that symposium. It constitutes the proceedings in the sense that the book is based on papers that were presented at the symposium. At the same time it is not an official record or account of what was said and done at the symposium for a variety of reasons. First, several speakers never submitted written manuscripts. More important, however, is that none of the papers in this book is in the same form in which it was delivered in Oosterbeek. I have edited all the papers, a few slightly, many heavily, and have completely rewritten several. In many cases, I have redesigned illustrations and had them redrawn. The responsibility for all these changes is mine. In every case I have, of course, submitted the edited or rewritten version of the paper to its author for his approval. In some cases, the process of editing and submitting manuscripts to authors had to go through more than one cycle. Despite these precautions, it is possible that some errors or misstatements managed to escape the critical eyes of author, editor, and copy editor. For any such errors, I myself must be held accountable.

DEPTH AND COVERAGE

When I first started planning the symposium and looking for specialists to discuss the various topics around which I had organized it, I suspected that I would find great variation in the amount of quantitative data available on those topics. My suspicions were entirely confirmed. Although the applied anthropologists who contributed to the symposium and to this volume complain about the paucity of data on the anthropometric dimensions of the various peoples of the world, they are, comparatively speaking, blessed with a wealth of data. By contrast, cross-cultural data on certain other aspects of human factors are virtually nonexistent. Readers of this volume may, in fact, find it interesting to read as an introduction to this book an article in which I specifically discuss some of the ergonomic topics on which we have little or no cross-cultural data.*

Whenever experts agreed that there were no data on a topic, I encouraged speakers and authors to present hypotheses, personal observations, case studies, and plans or needs for further research. As a result, the contributions to the symposium varied greatly in depth and coverage. The contents of this book reflects rather faithfully the character of the symposium. Some of the articles are packed with data, some have only a few data, others are frankly very thin. But that is the state of the field. This book, in short, says as much about what we do not know as about what we do know. My hope is that what I have put together here will serve as a stimulus for research as well as an introduction to a new facet of human factors engineering.

ACKNOWLEDGMENTS

Above all, I owe my thanks to the Scientific Affairs Division of the North Atlantic Treaty Organization for its support of the symposium and of this book. I sincerely hope that the book will prove the project worthy of the support that so many other agencies could not, or would not, give it.

Mrs. Doris Stude spent hours carefully deciphering my handwriting, notes, and scribbling on manuscripts prepared in every conceivable style.

*Chapanis, A. "National and cultural variables in ergonomics." *Ergonomics*, 1974, 17, 153–75.

Her meticulous work in typing and retyping every word in this book has helped to bring the project to a happy completion. Mrs. Margaret Iwata was responsible for all correspondence, other paperwork, and financial transactions. I owe more to both ladies than I can possibly put in words.

A few people, by custom and for other reasons, shall remain nameless. Although I have not acknowledged them individually, they will recognize their several influences throughout the book and will know intuitively how deeply grateful I am to them for the contributions they made in so many important ways.

ALPHONSE CHAPANIS

Some of the Participants in the

SYMPOSIUM ON NATIONAL AND CULTURAL VARIABLES IN HUMAN FACTORS ENGINEERING

Hotel de Bilderberg, Oosterbeek, The Netherlands,

19–23 June 1972

(1) George E. Rowland, (2) Chittranjan N. Daftuar, (3) Jacques Sporcq, (4) Paul K. Verhaegen, (5) Masamitsu Oshima, (6) Henry G. Maule, (7) Sadao Sugiyama, (8) H. Wallace Sinaiko, (9) Stanley Lippert, (10) Maria Wiersum-Jaeger, (11) Otto G. Edholm, (12) Anthony N. Nzeako, (13) Charles R. Brown, (14) Cyril H. Wyndham, (15) Enrica Fubini, (16) Alphonse Chapanis, (17) Vicki V. R. Cohen, (18) Alain L. Wisner, (19) Eckehard Behr, (20) Paul Branton, (21) Lewis F. Hanes, (22) John R. de Jong, (23) Thomas S. Fraser, (24) Richard S. Hirsch, (25) Anthony H. Marsh, (26) Erika Stief, (27) Baruch Givoni, (28) Max S. Schoeffler, (29) O. R. Spilling, (30) Wolfgang Weber, (31) H. P. Ruffell Smith, (32) Robert M. White, (33) Spyros E. Diamessis, (34) Ralph B. Archbold.

Introduction and Welcome

JOHN R. DE JONG

It is with great pleasure that I welcome you here at last after our lengthy preparations for this symposium, preparations in which Al Chapanis has had by far the greatest share.

If a history of Arbeitswissenschaft, or Ergonomics, or Human Factors Engineering, were to be written, the following four periods might be distinguished:

- The period in which man in general was the object of study.
- The period in which average human capacities and dimensions were determined and taken into account in adapting work to man. An example is the standard of 4.2 work kilocalories per minute, as the permissible physical work load.
- The period in which attention was increasingly paid to the spread or range of human capacities and dimensions.
- The period that is just beginning, in which differences among groups associated with cultural and national factors are accepted as important in research and work design.

I am afraid I cannot claim that a good reason for holding this symposium in this country is that we have paid particular attention to cultural and national variables. However, the Low Countries, and, of course, I include our Benelux partner, Belgium, have been aware of the diversity that marks

John R. de Jong, Raadgevend Bureau Ir. B. W. Berenschot N. V., Churchilllaan 11, Utrecht 2500, The Netherlands.

man as long as a hundred years ago. Quételet, at one time Director of the Royal Observatory in Brussels, mentions in his book *Soziale Physik* (1869) data regarding human capacities and points out that age and sex are relevant factors in determining those capacities. So the third period that I mentioned appears to have begun in this region more than a hundred years ago.

Eventually, of course, we shall also have to take into account that the diversity of human capacities and dimensions is itself changeable. An example is stature. One difficulty in interpreting the data is that maximum stature is reached at about twenty years of age nowadays and was reached at a somewhat older age in years past. Nonetheless, the trend is unmistakable: in this country, the mean stature of males has increased during the last decade by some 3 mm per annum (see Fig. 1). There is a still more striking example of inconstancy. The length of hair of Dutch males between the ages of fifteen and thirty years appears to have increased about 12 mm per annum during the last five years!

Some other examples of recent changes in this country are:

- The decreasing number of workers who work on shifts.

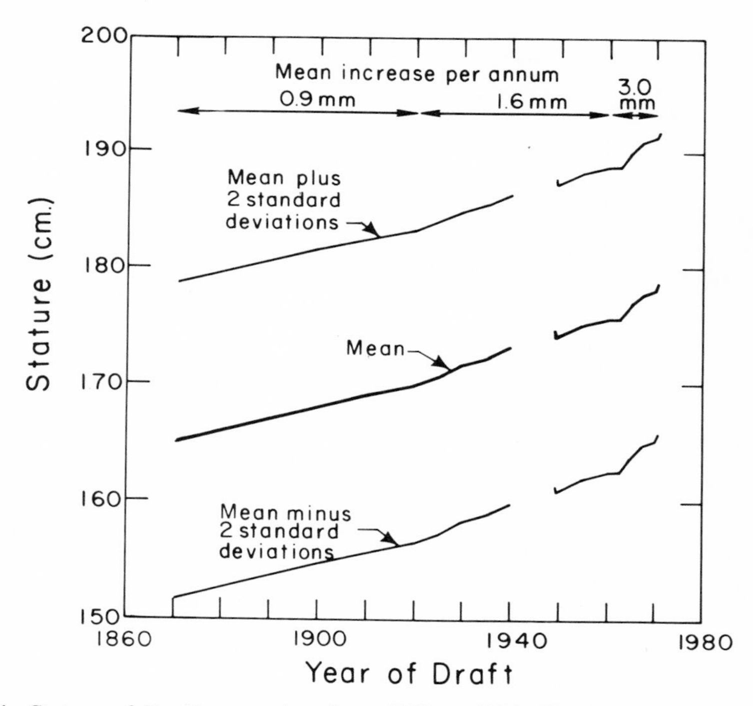

Fig. 1. Stature of Dutch conscripts from 1870 to 1971. The mean age of the conscripts from 1870 to 1934 was 19.5 years; from 1935 to 1971, 18.5 years. (*Source*: Centraal Bureau voor de Statistiek/Inspectie Militair Geneeskundige Dienst.)

- The almost certain repeal of the law that prohibits women from working night shifts in industry.
- The increasing popularity of job enrichment or job enlargement which includes the replacement of short-cycle, conveyor belt, assembly work by the assembly of larger production units by much smaller groups of workers without a conveyor belt.

That at least some attention is paid in The Netherlands to national and cultural variables will appear from the contributions of two Dutch participants. The topics of their papers are associated in part with the multinational character of their organizations, and in part with the circumstance that many workers from Southern Europe are now employed in this part of Europe. For all that, it is clear that we have much more to learn than we have to teach. The former I shall begin to do immediately.

I wish you a most fruitful and pleasant time in our country and at this symposium.

ETHNIC VARIABLES
IN HUMAN FACTORS
ENGINEERING

ONE

Cosmopolitanism: A New Era in the Evolution of Human Factors Engineering

ALPHONSE CHAPANIS

Human factors engineering, or its equivalent, human engineering, is a term used almost exclusively on the North American continent. Everywhere else, with the possible exception of the U.S.S.R., *ergonomics* is the term that most closely approximates its American counterpart. Although these two terms, *human factors engineering* and *ergonomics*, look and sound vastly different, the differences between the disciplines are more apparent than real. To a large extent, the invention and gradual acceptance of one or the other of these two terms in different regions of the world is largely the result of historical accident and the normal processes of cultural diffusion. For all practical purposes both human factors engineering and ergonomics have common aims, methodologies, and principles. Both disciplines apply information about human characteristics, capacities, and limitations to the design of human tasks, machines, machine systems, living spaces, and environments so that people can live, work, and play safely, comfortably, and efficiently.

I have emphasized the essential identity between human factors engineering and ergonomics because in this paper I talk exclusively about human factors engineering. That, however, is merely a reflection of my American heritage and upbringing: the term human factors engineering comes more naturally to my mind. The important point to remember is that whenever I write human factors engineering you may, if you like, substitute the word ergonomics with no loss of meaning.

Alphonse Chapanis, Department of Psychology, The Johns Hopkins University, Baltimore, Maryland 21218, U.S.A.

THE PURPOSE OF THIS BOOK

The purpose of this book is to bring together in one volume a collection of papers on national and cultural variables in human factors engineering. Nothing like this has ever been done before. I regard it not only as a unique collection of papers but as the threshold of a new era in the evolution of human factors engineering. It is this latter point that I want to elaborate in this paper.

STAGES IN THE DEVELOPMENT OF HUMAN FACTORS ENGINEERING

The history of any science or discipline can never be traced in distinct stages with clear lines of demarcation between them. History is more like a moving stream than a series of steps and platforms. Nonetheless, if we look at the history of human factors engineering we can discern several more or less definite stages in its evolution.

PRE-WORLD WAR II

The first stage in the history of human factors engineering is, I suppose you might say, the pretechnology period. I mean by that the period before human factors engineering had emerged as a separate technology with a name of its own. This period begins in some ancient time when man first began to fashion tools and implements for his own use. It stretches up to the beginning of World War II. You all know the early scientific history of our discipline. It was meager, spotty, and unsystematic. There was, of course, a lot of practical experience and folklore, some of it good (Drillis 1963; Oakley 1949), some of it bad. Then too there were the famous experiments by Gilbreth and Taylor at about the turn of the century, and a few scattered studies by people like Raymond Dodge, Knight Dunlap, and Carl E. Seashore during World War I. The years following World War I also saw the establishment of the Industrial Health Research Board and the National Institute of Industrial Psychology, both in England, and the Psychological Corporation in America. All three of these organizations occasionally sponsored studies that dealt with the redesign of work tasks and work environments. At best, however, that work can be characterized as sporadic and uncoordinated. Perhaps the only systematic development to occur in this period of pretechnology was the consolidation of the field of time-and-motion engineering.

THE WORLD WAR II AND AEROSPACE ERAS

The big impetus to the field of human factors engineering came in World War II, and I surely don't need to recount in great detail what happened at that time. The war needed and produced fearsome and complicated machines the likes of which man had never seen up to that time. These new machines—radar, sonar, high altitude and high speed aircraft, naval combat information centers, and air traffic control centers—placed demands upon their human operators that were often beyond the capabilities of human senses, brains, and muscles. Moreover, this new class of machines made demands on their human operators not so much in terms of their muscular power as on their sensory, perceptual, judgmental, and decision-making capabilities. These demands raised problems about human performance, capabilities, and limitations that could no longer be answered by common sense and the time-and-motion engineer's principles of motion economy.

So it was that during World War II a new group of scientific experts turned their efforts to the integration of man into the new and complicated machine systems that were the products of the war effort. These new recruits were not engineers but biological scientists—psychologists, physiologists, anthropometrists, and physicians. It was as a result of their influence that human factors engineering, as we know it today, emerged as a separate discipline. Immediately following World War II, the growth of human factors engineering was rapid, but still largely confined to machines of war and to closely related systems. It reached its climax in the aerospace industry and in the spectacular and breathtaking systems that have taken men a quarter of a million miles through outer space to our nearest celestial neighbor, the moon.

THE ERA OF SOCIOTECHNICAL SYSTEMS

Human factors engineering entered a new phase a few years ago, a phase in which it turned its attention to problems of contemporary society. It's difficult to say exactly when this new phase began, but it was a fairly noticeable transition. In 1965 everyone was still talking about military systems, aircraft, and space vehicles. Then suddenly in 1969 everyone seemed to be talking about different kinds of problems—problems of consumer goods, transportation, highways, urban design, health facilities, schools, housing, law enforcement, pollution control, airports, and urban power systems.

I don't mean to say that human factors engineering has turned away completely from military and aerospace problems, because that certainly is not the case. Many of our colleagues are still engaged in that kind of work. Nonetheless, there has been within the past half-dozen years a very distinct shift in the nature of the field, in the kinds of jobs that human factors engineers have taken, and in the kinds of papers that are reported at technical meetings and in our journals.

This shift in attention to the problems of contemporary society has been a wholesome one for the field. It has forced us, for example, to look at populations other than young healthy adult males. We learned that we had to design things for young children and old people, for the healthy and the infirm, and for women as well as men. At the same time, these new emphases have forced us to turn our attention to kitchen stoves, typewriters, ladders, dental equipment, houses, subways, and patrol cars for policemen. These new problems and solutions have broadened our perspectives, our horizons, and our prestige in the world around us.

THE COSMOPOLITAN ERA

The new era into which I see us now entering is even more exciting and likely to have an even greater impact on the field. For this I see as the beginning of the cosmopolitan era—the era in which human factors engineering loses its national parochialism and becomes a sophisticated discipline of the world.

I'm sure I don't need to remind you that up to the present time human factors engineering has been largely an American and Western European discipline. Until a few years ago almost all the research in the field had been done in America and Western Europe. Our findings were geared to large-boned peoples, peoples who were born to or used the English language, and peoples who had Western customs, habits, and ways of life. That's a pretty small percentage of the world's population. In the meantime, there had been some scattered reports, mostly in the anthropological literature, about problems that had been encountered when Western technology was introduced into the less developed countries of the world. In addition, various agencies of the United Nations have long been concerned about the problems of industrialization in underdeveloped countries. Indeed, the eminent anthropologist Margaret Mead edited a manual on some of these problems as long ago as 1953 (Mead 1953).

Even as I was preparing this talk an article came across my desk that I think is worth quoting from briefly. The author (Dart 1972), an American professor of physics, had spent some time in Nepal and had made a series of elementary anthropological observations about Nepalese ways of thinking. Here is one of his observations: "The predominant view pictures human knowledge about nature as a closed body, rarely if ever capable of extension, which is passed down from teacher to student and from generation to generation. Its source is in authority not in observation. In fact, experiment or observation was never directly suggested to us as an appropriate or trustworthy criterion of the validity of a statement nor as its source" (p. 53). Here is a people with a view about knowledge that is completely antithetical to the intensely practical, empirical approach of the human factors engineer.

Let me give you just one more quotation about a somewhat different kind of problem. In summarizing some observations about the use of pictorial representations, Professor Dart says, "The villagers use no other kind of map; they do not use drawings in constructing a building or a piece of furniture—in fact they hardly use drawings or spatial representations at all, and lack of spatial models is very natural" (p. 54). Imagine the difficulties a human factors engineer would have in talking about blueprints, models, or pictorial visual displays in such a culture.

But we do not even need to go to such extremes to find examples of vast and important differences among the peoples of the world. Consider, for example, the problem of language (Chapanis 1965). Americans and the British sometimes claim chauvinistically that English is becoming the language of the world. That may very well be, but it will be a long time before we can design equipment on the assumption that our customers will understand English, or that, if they understand it, they will be able to use it or to think in the English language. The fact of the matter is that our world contains a bewildering assortment of languages. No one really knows how many different tongues there are, but realistic estimates place that number at about 3,000! If we concentrate only on the major languages of the world, we find that 130 are used by more than 1,000,000 people each, and 70 by more than 5,000,000. Sixty percent of the people on the face of the earth speak some language other than the five official languages of the United Nations—English, French, Russian, Spanish, and Mandarin Chinese. Several papers later in this book (Sinaiko, Brown, Ruffell Smith, Hanes) remind us forcefully of problems that the human factors engineer faces when he tries to design equipment for non-English-speaking customers.

To continue with examples of this sort would be to steal from the many contributors to this book. The few that I have given you, however, are perhaps sufficient to convey to you why I am so excited about the new era into which I see us entering. With this book we at last recognize that we are one world and that a truly viable technology must have principles and practices robust enough to face up to the diversity we see around us and to design for man wherever we find him. This is a new level of sophistication for our discipline.

THE TIMELINESS OF THIS BOOK

In a sense, I think this book could not have been timed more appropriately. We are bringing these problems to the attention of the human factors world at exactly the time when the world is hungry for the information we have collected. Let me tell you just a few reasons why I think this is the case.

THE GROWTH OF HUMAN FACTORS AROUND THE WORLD

Human factors engineering is rapidly becoming a truly world-wide discipline. Although America and Britain may still dominate the field, it has at least outgrown their borders. The International Ergonomics Association (IEA) now contains over 3,000 members distributed throughout 30 countries of the world. It contains 12 member societies and, if we exclude the American Human Factors Society and the British Ergonomics Research Society, the remaining 10 societies contain more than 1,300 members. These very impressive figures remind us that we have indeed become world-wide. The more important implication, however, is that if our members are scattered over the wide reaches of this globe then our technology should be broad enough to provide the principles and practices that can be applied by our colleagues no matter where they may be.

THE NEED FOR INTERNATIONAL STANDARDS

Not only has the IEA been growing but it has also begun to take a more active role in spreading human factors throughout the world. One significant step in that direction is the organization, under the sponsorship of the IEA and the International Standards Organisation, of an International Symposium on Ergonomics and Standards that took place early in 1973. That meeting was designed to bring together human factors experts from many countries to explore the problems and to develop the procedures for preparing international human factors (or ergonomic) standards for the design of equipment. I regard this book as a necessary prelude to that important work. We cannot have international standards until we understand the influence of national and cultural variables on the principles and practice of human factors engineering. The preparation of standards is, of course, a job that is not done once and then forgotten. Standards must be revised in the light of new experiences and data. As more and more findings about national and cultural variables become available, we can hope to see greatly improved standards that will benefit all of us in our work.

OUR WORLD HAS SHRUNK

The third reason why I think this book is so timely is that our world has shrunk greatly in the past few centuries. You understand, of course, that I don't mean that our physical earth has shriveled up literally. I do mean that we are all much closer together now than we have ever been. Let me remind you of just a couple of milestones in the history of our civilization. When Ferdinand Magellan began the first recorded circumnavigation of our globe he left the port of Salúcar de Barrameda, Spain, on September 20, 1519. Magellan himself never completed the trip, but his ship, the Victoria, returned to Spain on September 8, 1522, after a voyage of almost exactly three years. Four hundred years later, the first aerial circumnavigation of the globe took place in 1924 when three U.S. Army planes took off from Seattle, Washington, on April 6, and returned to the same airfield on September 28. That was a trip of almost six months. Today, less than fifty years later, a commercial airliner travels around the earth in approximately sixty-seven hours, and that time includes all its normal stops at a dozen or so cities. Of course, if you are an astronaut you can make the same trip in about ninety minutes!

However impressive and symbolic such comparisons may be, their consequences are even more impressive. Today Americans buy automobiles, typewriters, tape recorders, and calculating machines made in France, Germany, Britain, and Japan. We, in turn, sell our computers, aircraft, and farm machinery to Europe, Africa, South America, Western Europe, and perhaps soon to Communist China. Similarly, the European Common Market has resulted in a free interchange, not only of products, but of workers among the several countries belonging to that confederation.

These are only a couple of examples of the internationalization of business and commerce that has become an important fact of modern life. World trade rose an average of 8 percent per year during the 1960s—a growth rate considerably higher than that experienced by many domestic economies. By 1980 the volume of world trade is expected to reach $1,000 billion in U.S. dollars. Direct U.S. foreign investments increased from $12 billion in 1950 to $55 billion at the end of 1966—an average annual growth rate of 10 percent. They will total more than $200 billion in current dollars by the end of 1980. British private overseas investments have also grown at about 10 percent a year since 1962, and comparable figures from the Federal Republic of Germany have shown an average annual rate of 18.8 percent in recent years.

Coupled with this growth in international trade has been an important change in the philosophy of international marketing. Until a few decades ago, the approach was to handle markets geographically in such regional divisions as Latin America, Europe, North America, Australia and the Pacific, and Asia. Within each region, products were individually designed, manufactured locally, and distributed in the same regions where they were produced. In recent years, a completely different philosophy seems to be taking over. The latter emphasizes an integrated approach to multinational marketing wherein products are standardized world-wide and sold as such around the world. The advantages of such standardization are in such things as reduced costs of production, greater consistency in products and in services to customers, and increased efficiency in sales and marketing.

The new trend has not been without its problems. The Stanford Research Institute (1970) recently compiled an impressive collection of extremely costly marketing failures that could be traced to ignorance of behavioral factors. It is important to add that the failures were not all on the part of American manufacturers. The list covers food products, automobiles, home appliances, beauty products, recreational equipment, and home furniture. These are costly lessons to learn and I suspect that manufacturers the world around are going to be increasingly eager for whatever information we can give them about national and cultural variables in the design and marketing of the products they produce. I see this book as a first step to producing the kind of information that they need and that I hope we will eventually be able to supply.

THE NEEDS FOR TECHNOLOGY IN THE UNDERDEVELOPED AREAS OF THE WORLD

The final reason why I think this book is so timely is that we must respond to the needs of underdeveloped countries. These countries are almost literally crying for modern technology. Gross discrepancies that we find among the nations of the world are, in present-day morality, outrageous. They cannot be allowed to persist. In 1965 the developed countries of the world contained a little more than 21 percent of the population, but nearly three-quarters of the gross national product of the world (72.6 percent). These figures are certainly going to change. Indeed, United Nations sources predict that by 1980 the percentage of the world's gross national product contributed by the developed countries will have decreased to 64.9 percent (see Fig. 1). Evidence that such changes are already happening is not hard to find. For years now the developing countries have had a much higher average annual rate of growth in manufacturing, rail traffic, and electric energy consumption than have the industrialized countries (see Table 1). The needs are there, they are being filled, and the trend is almost certain to continue in the decades to come.

Although the underdeveloped countries greatly need the benefits of modern technology, the fact is that they themselves cannot fill that need. They have to depend on the products, the methods, and the technical skills of their more advanced neighbors. The less developed countries have neither the trained manpower nor the knowledge to do the job themselves. And that, of course, is where we come into the picture again. We have an unusual opportunity to introduce modern technologies, agricultural methods, manufacturing methods, systems of communication, and systems of transport in places where they have sometimes never existed before. Here is an opportunity for us to do a first-rate job of fitting modern technology to man and to do it properly from the start. How well we do that will depend to a large extent on the breadth of our vision and the universality of our principles.

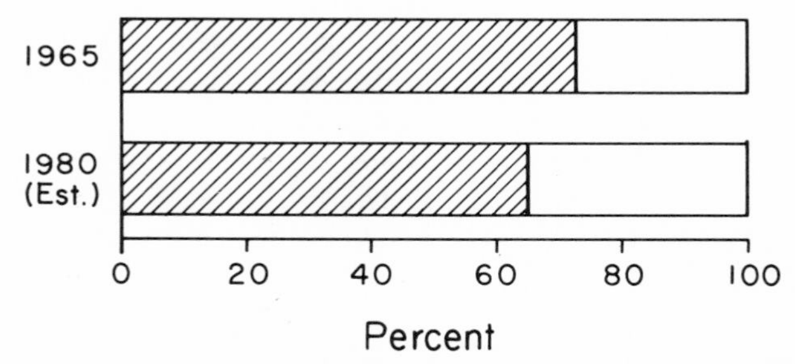

Fig. 1. Percentage of the world's gross national product accounted for by the developed countries (*striped areas of the bars*) and the less developed countries (*open areas of the bars*) of the world in 1965 and 1980 (estimated). (*Source*: United Nations.)

Table 1. Average annual rates of economic growth in the developing countries and the industrialized countries of the world[a]

	Developing countries (%)	Industrialized countries (%)
Agricultural production, 1960–66	2.1	1.8
Manufacturing output, 1960–67	7.3	5.6
Railway traffic, 1948–67		
Freight	4.4	1.1
Passenger	2.9	1.8
Electric energy production, 1948–67	10.5	7.7
Steel consumption, 1950/52–67	8.1	4.5

[a] *Source*: Commission on International Development, 1969.

CONCLUSION

This book is only a beginning, but it is an important beginning. It represents, to my way of thinking, the beginning of a new stage in the evolution of human factors engineering. Here for the first time, we recognize the impact of national and cultural variables in human factors engineering. From these beginnings I expect to see the development of a truly cosmopolitan technology, a technology that will be able to design for man so that we can all live and work more harmoniously with each other and with this system we call our earth.

REFERENCES

Chapanis, A. Words, words, words. *Human Factors*, 1965, **7**, 1–17.

Commission on International Development. *Partners in development*. New York: Praeger, 1969.

Dart, F. E. Science and the worldview. *Physics Today*, 1972, **25** (6), 48–54.

Drillis, R. J. Folk norms and biomechanics. *Human Factors*, 1963, **5**, 427–41.

Long Range Planning Service. Behavioral factors in world consumer markets. Stanford Research Institute Report No. 409. Menlo Park, California: July 1970.

Mead, M. (Ed.) *Cultural patterns and technical change*. Paris: United Nations Educational, Scientific and Cultural Organization, 1953.

Oakley, K. P. *Man the tool-maker*. Chicago: University of Chicago Press, 1949.

TWO

Population Differences in Dimensions, Their Genetic Basis and Their Relevance to Practical Problems of Design

D. F. ROBERTS

The metrical characterization of individual differences in size and shape derives, via the pioneer work of the Belgian astronomer Quetelet who published his *Anthropometrie* in 1870, from almost the earliest of artists who attempted to establish the rules of proportion or "canons" of the human form. In ancient Egypt, the length of the middle finger was taken as a third of the height of the head and neck, and a nineteenth of the whole body. Vitruvius in Roman times, Leonardo da Vinci, and Albrecht Dürer followed the rule that the head from crown to chin is contained eight times in the stature.

In the last century a vast body of anthropometric literature has been built up, and details of some body measurement are available for over ten thousand samples from populations in different parts of the world. These, of course, were not obtained for artistic purposes nor for purposes of equipment design, the subject of this discussion. Many studies were prompted by taxonomy, seeking affinities and differences among groups; by physiology, attempting to explain variation in body function; by clinicians interested in departures from normality; and by students of growth interested in how the body comes to be as it is.

The ergonomic implications of human variability were not fully realized at first. Though indeed there were several classic studies in the 1920s and

D. F. Roberts, Department of Human Genetics, University of Newcastle upon Tyne, Great Britain.

30s, it was not until World War II that maximization of efficiency of operators of precision machinery was seriously attempted by correct positioning of controls, of seating, and of body supports to meet the dimensional and physiological requirements of the user. Since then, a wider application of the principles to a great variety of domestic and industrial situations has come about (Le Gros Clark 1949; Roberts 1962, 1964; Ministry of Housing 1962). Right from those early days of ergonomics, long before the discipline received its name, came the recognition that individuals within a population vary, and that it was necessary to design machinery, equipment, and environments to accommodate not only the average user but the greater part of the range of variation that describes all users.

Only recently, however, have the problems of variation between populations begun to emerge. Two factors in particular appear to be responsible for these newer and broader concerns. The first is the enlargement of world trade in sophisticated products. The fundamental problems of dimensionally suiting either population *X* or population *Y* are very similar, and to design for one instead of the other entails substituting the essential measurements, if these are available. So, for example, anthropometric data have recently been collected on German and Vietnamese forest workers (Kurth & Lam 1969) to help in the design of German tools and equipment for use in Vietnam. To design for two or more populations together may be less easy. Recently, for example, dimensions were submitted to the European Standards Co-ordinating Committee, Working Group 43 "Office furniture," of the International Standards Organisation (Table 1). Table 1 shows that neither the sitting heights nor standing heights of the three populations differ excessively. Knee heights (popliteal-floor distances), however, pose a problem. The mean knee-height for German males exceeds the 95th percentile of both British and French males. If information had been collated from all populations, would the problem have been greater still?

Table 1. Anthropometric dimensions (in mm) submitted for consideration by the International Standards Organisation

		United Kingdom			France			Germany		
			Percentiles			Percentiles			Percentiles	
Dimensions	Subjects	Means	5%	95%	Means	5%	95%	Means	5%	95%
Size standing	Men	1,753	1,644	1,861	1,711	1,614	1,829	1,755	1,636	1,880
	Women	1,626	1,517	1,734	1,578	1,499	1,670	1,648	1,534	1,762
Size sitting	Men	919	860	977	879	831	944	925	861	988
	Women	854	796	912	827	782	868	843	808	875
Popliteal-floor	Men	430	396	464	419	387	453	485	451	507
(without shoes)	Women	398	364	432	385	355	416	396	377	419

A second factor is the internationalization of standards. So long as national standards bodies laid down specifications only for their own nationals, they encountered only problems of variation among individuals in a particular population. But as soon as representatives from several countries tried to establish common standards, they quickly became aware of differences among populations, not only in dimensions but also in customs, practices, and methods of using products.

There are, I think, four points to be considered:

1. How great are population differences in size and shape?
2. Are these differences genetic?
3. Are they important for design purposes?
4. Where they are important, can they be solved?

POPULATION DIFFERENCES IN SIZE AND SHAPE

I do not propose to set out long lists of population dimensions, nor do I propose to repeat the discussions of other authors on the amount of overlap of distributions of particular dimensions. Some of the latter, incidentally, may be somewhat pessimistic, especially when they are based on military data which are frequently truncated, leading to distortions both of mean and variability. Instead, consider first the extreme range of variation. For this purpose I have drawn on my file of several thousand samples, which have been used for a variety of purposes of both an academic and applied nature.

THE EXTREMES OF VARIATION AMONG THE PEOPLES OF THE WORLD

The smallest people on record is, of course, the pygmy people of Central Africa. Gusinde (1948) found a mean stature of 143.8 cm for 386 Efe and Basua adult males, and 137.2 cm for 263 females. Data on the related Aka show mean statures of 144.4 cm for 115 males and 136.7 cm for 110 females. Other pygmy groups outside Africa have mean statures several centimeters higher. The standard deviations for all pygmy groups average some 7 cm.

The tallest peoples for whom I have records are the Northern Nilotes of the southern Sudan; for a sample of Ageir Dinka, the mean is 182.9 cm, with a standard deviation of 6.1 cm; for a sample of 227 Ruweng Dinka, the mean is 181.3 cm, with a standard deviation of 6.4 cm. The mean stature of the Ageir Dinka women is 168.9 cm, with a standard deviation of 5.8 cm. The greatest mean is only 127 percent of the smallest for males, and only 123 percent for females. The tallest normal individual Dinka of whom I have a record was 210 cm. The smallest pygmy male was just under 130 cm. So the tallest normal individual recorded is only 162 percent of the smallest.

INTER- VERSUS INTRA-POPULATION DIFFERENCES

The mean stature for some 1,200 samples from Africa and Europe is 167.1 cm, and the standard deviation of the means is approximately 5.6 cm. The standard deviation of stature is known for 200 of these samples. The mean standard deviation is 6.1 cm. In other words, variation in total body size between populations is, if anything, less than that within populations. Tildesley (1950) examined a number of anthropometric dimensions in a similar way for indigenous populations all over the world. Table 2 summarizes her findings. For brevity, I have made some omissions, mainly of head and facial dimensions. In almost every dimension, the variability within populations was greater than that between populations.

POPULATION DIFFERENCES AND CLIMATE

So much for the magnitude of population differences. The second point to make is that population differences are not random. When one examines the data on a world scale, a clear pattern emerges. Longitudinal limb characters, arm length and leg length, tend to be greater in proportion to overall size in peoples living in warmer climates (Fig. 1). Transverse and antero-posterior body dimensions, limb and trunk girths all tend to be greater, again relative to size, in peoples living in cold environments. In other words, those dimensions making for linearity of body form tend to be associated with warmer climates, those making for sphericity of form and increased mass tend to be associated with colder climates. The relevance of this relationship for climatic adaptation has been discussed on a number of occasions. Such variations in body shape facilitate the maintenance of body heat balance in different climates (Roberts 1952, 1972). How these differences come about during the growth of the individual and how they relate to nutrition and other environmental features have also been investigated. The most reasonable explanation is that these differences are selectively advantageous in relation to climate. That is to say, they have some genetic basis that has been selected for over many generations.

THE GENETIC BASIS OF POPULATION DIFFERENCES

To what extent are population differences genetic? Most of the evidence regarding the inheritance of size and shape is indirect. First, there is the limited evidence from groups growing up in environments to which their forebears were not indigenous. Such studies have been done, for example, on Europeans in New York, Japanese in Hawaii, Chinese and Mexicans in the United States, and on Japanese and Italian-Swiss on the Pacific coast of the United States. Some of these groups have been resident in their new environments for several generations, others are newly mi-

Table 2. Dimensional variation among and within populations (all measurements in mm)

Dimension	Population means			Population standard deviations		
	Number of populations	Mean of population means	Standard deviation of population means, i.e., interpopulation standard deviation σ_b	Number of populations	Mean standard deviation of populations, i.e., intra-population standard deviation $\overline{\sigma}_w$	$\frac{\overline{\sigma}_w}{\sigma_b}$
Nose-breadth	370	37.1	3.71	217	2.87	0.774
Sitting height	266	864.3	31.93	129	33.44	1.047
Nose-height (to nasion)	255	53.4	3.37	177	3.82	1.134
Bizygomatic breadth	402	139.2	4.55	245	5.23	1.150
Stature	573	1652.0	49.75	296	58.87	1.183
Span	147	1740.0	59.85	102	71.81	1.200
Morphological face-height (to nasion)	249	120.5	4.91	208	6.42	1.306
Hand-breadth (direct)	65	87.4	6.19	41	4.47	0.722
Shoulder-breadth (bi-acromial)	94	370.2	19.02	80	18.87	0.992
Mouth-width	82	54.3	3.61	34	3.73	1.034
Arm-length (projective)	63	742.5	25.12	33	32.90	1.310
Chest-girth (at rest)	76	887.9	38.02	39	49.89	1.311
Head-girth	97	554.1	10.64	56	14.45	1.357
Pelvic-breadth (ilio-cristal)	71	280.3	11.63	55	16.33	1.403
Breadth between inner eye corners	104	33.1	1.62	44	2.67	1.650
Physiognomic face-height	101	184.3	5.32	46	8.94	1.681

grant. All such studies demonstrate uniformly a tendency for the stature and other longitudinal body dimensions of the migrants to be greater in the new environment as compared with those of their kin remaining at home. These studies clearly suggest the occurrence, if not the extent, of human plasticity, perhaps through some direct influence on growth. Other possible explanations are that the migrant subgroups may be different from their parent stock, or that those with particular growth patterns tend to choose the habitat to which they are most suited. Changes in mating patterns giving increased heterozygosity and heterosis may also be partly responsible.

The relative stability of body proportions, on the other hand, is quite well documented. Several studies in the southern United States show clear differences between the body proportions of negroes and whites living under similar climatic conditions, with hybrids showing intermediate values. Shapiro's (1939) analysis of Japanese migrants in Hawaii demonstrated "that although the Hawaiian-born are taller and have larger sitting

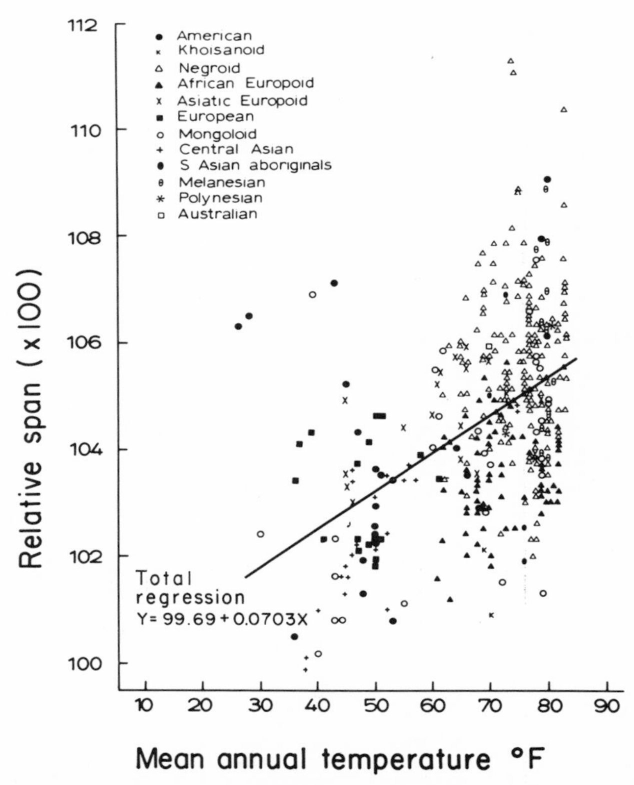

Fig. 1. Relative span (maximum distance from fingertip to fingertip as a ratio of stature) for 372 samples plotted against mean annual temperatures.

heights than the immigrants, nevertheless the proportion remains unaltered." Thus populations may tend to retain their characteristically individual relative sitting heights in different environments, with the result that different populations living side by side under identical conditions may have distinctly different proportions. The data suggest a strong genetic component to body proportion, and a more labile overall size.

GENETIC BASIS

There is relatively little direct genetic evidence on the way body dimensions are inherited. A few single genes are known to produce a pronounced effect on height, for example, achondroplasia and Marfan's syndrome, and height and body proportions are also affected in chromosomal conditions such as Down's syndrome. Apart from such clinical conditions, normal variations in height seem to be most reasonably explained on a polygenic hypothesis. Polygenes are transmitted in the same way as, and in accordance with the same laws as, major genes, but their effects do not provide sufficient discontinuity for individual study. A polygene acts as one of a system, the members of which may act together to effect large phenotypic differences, or against each other so that similar phenotypes develop from different genotypes. An individual polygene has only a slight effect; it is apparently interchangeable with others within the system; it does not have an unconditional advantage over its allele, since its advantage depends on other alleles present in the system; and it cannot therefore be heavily selected for or against (that is, not until its associates in the same system come to be linked with it).

In polygenic inheritance, similar phenotypes develop from different genotypes, and therefore great genetic diversity and potentiality for change is concealed behind the phenotypic variation. The result in the individual depends more on the numbers of genes for tallness that he carries than on the particular genes present. Figure 2 shows the variation that would be produced in a size character if three pairs of allelic genes were responsible for it, the gene for taller size being denoted by the capital letter in each case. The diagram assumes that the genes are all equal in their effects, that they simply add to each other, and that the alleles of each pair are equally frequent. With more genes, the irregularities of the steps of the histogram disappear, and the curve becomes smooth. Thus a normal curve of distribution of a quantitative character in a population may be attained entirely by its genetic determination. If such a population were uniformly exposed to a different environment increasing the character, the whole curve would shift to the right. If only part of the population were so exposed, the variability of the curve would increase (Fig. 3).

The extent of the genetic contribution to such quantitatively varying characters can be measured by partitioning the variance. In essence, the total phenotypic variance is divided into that due to (1) additive polygenic

THREE LOCI, TWO ALLELES AT EACH, EQUAL AND ADDITIVE IN EFFECT:
DISTRIBUTION OF GENOTYPES

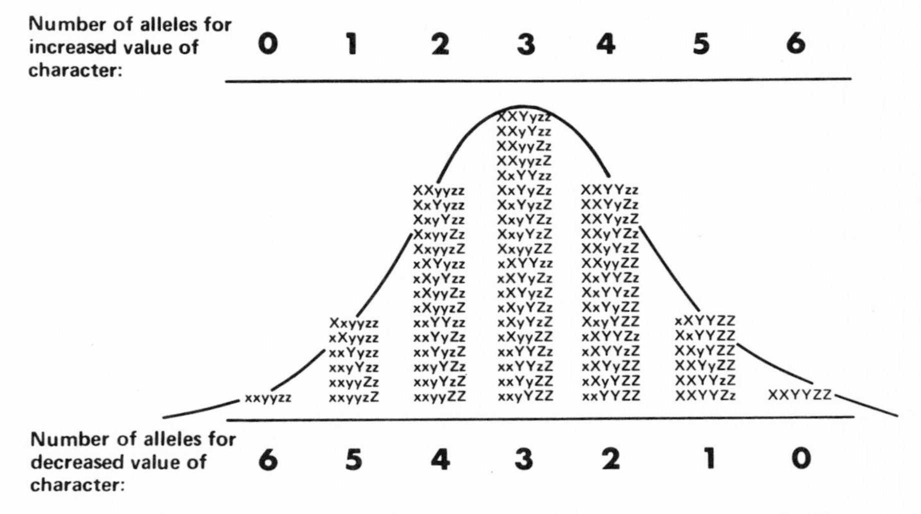

Fig. 2. Distribution of genotypes for three loci with two alleles at each. The example assumes that loci and alleles are equal and additive in effect.

effects, (2) environmental effects, and (3) other sources. The additive genetic contribution expressed as a proportion of the total variance is known as the heritability. A character that is totally genetically determined would have a heritability of 100 percent; that for which all the variation was environmental would have a heritability of zero. Heritability can be measured in a number of ways, for example, from the examination of resemblance between relatives of different degrees, and from twin studies. For stature, all the evidence points to a heritability of about 85 percent in Western European and African populations. But even if one establishes the genetic contribution to variation within a population as being, say, 100 percent, this implies nothing as to the genetic basis of variation between populations. Theoretically it is possible for variation within a population to be totally genetic and for variation between populations to be purely environmental.

At the moment, it is not possible, therefore, to quantify the genetic contribution to any dimensional difference between populations. All we can say is that the probability is very high that there is an appreciable genetic contribution to the striking differences among peoples. The variation in height between the tall Nilotes and the small pygmies must derive essentially from genetic differences. A sample of Welsh or Eskimo infants brought up in the Southern Sudan as Nilotes would certainly not develop

Nilotic physiques. But beyond that we cannot say. A genetic basis to population differences in size, i.e., differing frequencies of genes for tallness and shortness, appears reasonable when one remembers the considerable differences in gene frequencies that are well established for human populations, i.e., for monogenic characters such as blood groups and isoenzymes.

DESIGN IMPORTANCE OF POPULATION DIFFERENCES

It is difficult to generalize about the importance of these population differences in the application of dimensions to problems of design. Opposing views have been expressed, but it seems to me that these views are capable of resolution if an ergonomic rather than a philosophical approach is made to the problem.

DESIGNING FOR AN OUTSIDE LIMIT

The simplest problem is that in which a single dimension is required as an outside limit (Roberts 1956) for all or nearly all members of the populations being designed for. Here population differences need be of little practical importance. If one is accommodating the largest male in some one population, a slight increase will accommodate the largest in any population. If no such adjustment is made, however, discomfort can arise. Those

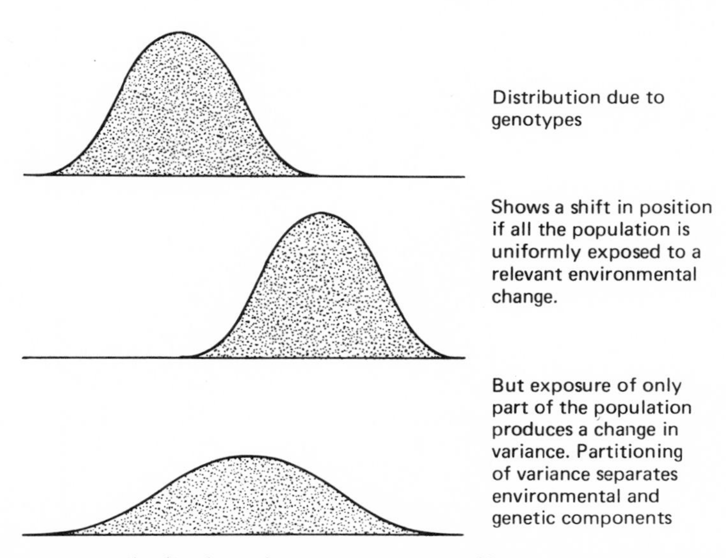

Fig. 3. Assumed distribution of genotypes (*top*). Shift produced by exposure of the entire population to a relevant environmental change (*center*). Distribution of characteristics when only part of the population has been exposed to the environmental change (*bottom*).

of us who have had the opportunity to use a domestic modern bath designed in Japan for Japanese or who have sat for several hours on end in a Japanese theater, will have experienced the difficulty of fitting even a moderate sized person from a large population into dimensions designed for a smaller population.

DESIGNING A SINGLE DIMENSION FOR EFFICIENT USE

Rather less straightforward is the recommendation of a single dimension for efficient usage. In many pieces of apparatus, efficient usage requires a narrower tolerance limit in one direction than in the other, and such limits may be imposed by only one or two of the operations for which the apparatus is required. Efficient usage by the majority of one population may be outside the limits of tolerance for many people from another population. The more critical the limits, the more necessary it may be to consider population differences. It is, moreover, not only important to consider population differences in size but also to consider population differences in modes of utilization of equipment.

DESIGNING MULTIPLE DIMENSIONS

Rarely is a designer able to worry about only a single dimension. The more typical situation requires a combination of dimensions for both limiting and working purposes, and here we begin to encounter the problems of different proportions as well as those of different size. These problems may vary in severity from the relatively simple to the highly complex. Many Europeans find the manipulation of small controls on Japanese instruments difficult; Japanese secretaries were not well fitted by the old style office furniture in use in the West; and, as Kennedy illustrates in his paper in this volume, the intricacy of an American cockpit is well-nigh impossible for optimal utilization by Japanese and Thai pilots. Thus the more complex the application, the greater the need for special surveys of and fitting trials for the different populations for whom the equipment is designed. But even quite intricate problems, on which interpopulation variation seems at first insuperable, may be capable of relatively simple solution. Two examples follow.

SOLUTION OF PROBLEMS OF SIZE DIFFERENCES

EXAMPLE 1: THE DESIGN OF KEYBOARDS

The design of keyboards is an example of an intricate ergonomic problem arising from critical variation among populations. Some years ago, we were concerned with the establishment of the metrical relationships of the fingertips when the hand is at rest in a position to use a keyboard. The extensive literature on the anthropometry of the hand that had accumulated since the first studies by Buffon in 1766 related almost entirely to descriptive dimensions. It referred almost entirely to the hand in a static position, usually with hand flat and fingers together. There had been, to be sure, a few studies of the hand in working or grasp positions to provide data for the design of cockpit and other controls. But we could find no data on how the fingertips lie in relation to each other when the hand is comfortably placed, at rest, in a position convenient for the operation of a keyboard.

We undertook an investigation to inquire whether there is a position of functional rest of the hand, and, if so, to identify it. We defined this position of functional rest as the position in which there was the most complete relaxation of the muscles and ligaments. The position of rest defined by Wood Jones (1920) was unsuitable for a functional position, since his definition requires the fingers to lie against one another and the distal phalanges of digits 3, 4, and 5 to be bent under the other phalanges so that any movement of one finger is impeded by the others. An obvious restriction of our definition was that the lower arm must be pronated to bring the hand into a position in which the tips of the fingers could be placed over the keys. When the hand was resting thus on a flat surface, the rest position was defined as the position of the fingers and wrist in which there is minimal activity of the muscles of the hand and forearm.

In our study, this rest position was identified by electromyography and described by photogrammetry. A pair of silver disc electrodes 1 cm in diameter and about 6 cm apart was fastened by adhesive plaster to the skin over the muscle or muscle group to be studied and connected to the input grids of an Ediswan AC pushpull amplifier. A third disc electrode, connected to the earth terminal of the amplifier, was attached to the skin of the anterior aspect of the wrist where there are very few muscle fibers present. Additional pairs of electrodes were placed over other muscle groups to be studied. The first pair was situated on the lateral aspect of the palmar surface of the hand, one placed on the thenar eminence and one on the pad at the base of the index finger. In the second pair, one was placed on the hypothenar eminence and one on the pad at the base of the 5th finger. The third pair was situated on the postero-lateral aspect of the forearm over the main extensor group of muscles just below the lateral condyle of the humerus. Others were placed over various other groups. In all, four to six electromyograms could be recorded simultaneously on a multichannel pen recorder. The electrodes were positioned to give us readings from functional aggregates rather than recordings from specific muscles. Our intent, in short, was to obtain a general view of muscular activity in the small muscles of the hand and the extensors of the wrist and fingers.

Each of 30 male and 30 female subjects, seated on a chair 43 cm high at a table 71 cm high, was asked to place his right forearm on the table surface in a position that he felt to be comfortable, with his hand in a position that he thought was relaxed. The forearm, the heel of the hand, and the fingertips rested on the table throughout. He was then instructed to curl his fingers slowly until they were about three-quarters flexed, then to extend them fully. The hand was held motionless for a short time at these positions and at a number of intermediate positions as well. This entire cycle was repeated several times and electromyographic records were obtained throughout. It was a relatively simple matter to identify the position of minimal muscular activity (Fig. 4). Indeed in some subjects there was no activity at all in the relaxed position of the hand. When the hand was in this position of minimal activity, two simultaneous photographs were obtained, one from the vertical and one from the horizontal elevation. This standardized photographic technique, using long focus lenses, had been applied successfully in a number of investigations (Roberts 1956, 1959). From the photographs, measurements of the hand in plan and in side elevation were obtained. The positions of the fingertips in the horizontal plane were measured by rectangular coordinates, the fingertip being taken as the point of contact of the tangent to the most distal point of the fingertip. The line joining the tips of the styloid processes of the ulna and radius was chosen as the base line, and measurements were made from the reference datum of the midpoint of this line.

In thirteen of the men and eleven of the women this procedure of determination of the relaxed position by electromyography and photogrammetry was repeated ten times to determine intraindividual variability in the relaxed position. The position of rest of the hand is easy to identify and is

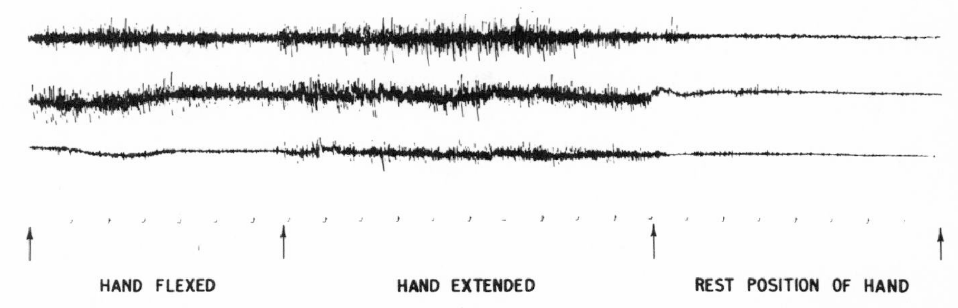

Fig. 4. Electromyographic recordings of the hand and lower arm with the hand in various positions. Extensors of lower arm (*upper trace*), lateral anterior muscles of the hand (*middle trace*), medial anterior muscles of the hand (*lower trace*). This segment of a record is about 25 seconds long. The time trace (*bottom*) marks off one-second intervals.

remarkably reproducible. The average intraindividual variance in fingertip position obtained from the repeated poses both in males and females is very slight (Table 3). In a given individual, there is very little variation in the position of the digit in the longitudinal plane, and somewhat greater variation in the transverse plane. This reproducibility, particularly striking since the hands were removed completely from the table between poses, indicates that for a given individual a position of rest of the hands exists and that it can be identified. Further, analyses of variance (Table 4) show that the position of rest is a highly individual characteristic, presumably related to slight differences among individuals in the musculature of the hand and forearm. Measurements were also obtained from one pose for each of the subjects, and the means and variances of these measures are given in Table 5.

Despite the individuality, as shown by the highly significant mean squares between subjects (Table 4), and the significant differences between mean positions for males and females (Table 5), the design problem can be simply resolved. The fingertip positions of digits 2 to 5 in females and males are practically superposable if the female hand is placed a little farther forward on the table to allow for the smaller wrist-fingertip distance (Fig. 5). When so superposed, the thumb positions in males and females are side by side. This finding indicated that a keyboard based on these resting fingertip positions would be equally usable by both sexes if the keyboard were correctly positioned for the hand and if the thumb key were designed as a bar instead of a round button.

The individuality and criticality of finger position when the hand is at rest are likely to occur in all populations, and the position of rest is likely to differ appreciably among them. A great deal of information shows that

Table 3. Intraindividual variability (average variances for trials within subjects) in the relaxed fingertip positions for 11 females and 13 males (all measures in cm)

		Females	Males
Finger 1 (thumb)	Transverse	0.360	0.457
	Longitudinal	0.038	0.084
Finger 2	Transverse	0.352	0.330
	Longitudinal	0.101	0.097
Finger 3	Transverse	0.477	0.369
	Longitudinal	0.130	0.083
Finger 4	Transverse	0.277	0.372
	Longitudinal	0.116	0.072
Finger 5	Transverse	0.330	0.260
	Longitudinal	0.045	0.082

Table 4. Analyses of the variance between and within subjects for fingertip positions of the relaxed hand, for 13 males and 11 females (data in cm)

	Males						Females					
	Transverse			Longitudinal			Transverse			Longitudinal		
Source of variation	Degrees of freedom	Mean square	F	Degrees of freedom	Mean square	F	Degrees of freedom	Mean square	F	Degrees of freedom	Mean square	F
Finger 1												
Between subjects	12	10.202	22.34	12	6.855	81.40	10	15.305	42.53	10	5.038	132.14
Within subjects	116	0.457		116	0.084		98	0.360		98	0.038	
Finger 2												
Between subjects	12	29.947	90.81	12	11.780	121.65	10	11.299	31.77	10	14.401	141.27
Within subjects	111	0.330		111	0.097		97	0.356		97	0.102	
Finger 3												
Between subjects	12	59.820	162.02	12	10.809	130.44	10	17.071	35.42	10	11.635	88.33
Within subjects	111	0.369		111	0.083		98	0.482		98	0.132	
Finger 4												
Between subjects	12	58.767	157.81	12	10.517	146.62	10	12.067	43.51	10	10.312	89.02
Within subjects	111	0.372		111	0.072		94	0.277		94	0.116	
Finger 5												
Between subjects	12	55.085	211.67	12	2.871	35.02	10	14.498	34.20	10	16.834	377.98
Within subjects	110	0.260		110	0.082		89	0.424		89	0.045	

Note: The degrees of freedom for "within subjects" vary because it was not always possible to get 10 replications for every subject.

Table 5. Fingertip positions and wrist diameters of 30 males and 30 females with hand in the functional rest position[a] (measures in cm)

		Mean	Variance
Males			
Finger 1 (thumb)	Transverse	6.01	3.08
	Longitudinal	12.83	1.10
Finger 2	Transverse	0.88	4.11
	Longitudinal	17.36	1.43
Finger 3	Transverse	-1.92	5.47
	Longitudinal	18.07	1.69
Finger 4	Transverse	-3.70	5.13
	Longitudinal	17.06	1.36
Finger 5	Transverse	-5.86	3.65
	Longitudinal	14.31	0.96
Wrist diameter		6.41	0.24
Females			
Finger 1 (thumb)	Transverse	4.93	1.24
	Longitudinal	11.75	0.91
Finger 2	Transverse	1.30	2.74
	Longitudinal	16.22	0.99
Finger 3	Transverse	-1.07	3.96
	Longitudinal	16.88	0.93
Finger 4	Transverse	-2.67	3.22
	Longitudinal	15.78	0.77
Finger 5	Transverse	-4.71	3.00
	Longitudinal	13.22	0.87
Wrist diameter		5.83	0.06

[a]Fingertip positions on the ulnar side of the reference line are shown as negative deviations.

there are population differences in muscling of the hand and forearm. For example, the flexor carpi radialis has a common insertion with the palmaris longus, the lengths of which differ appreciably in African, Asian, and European peoples. The palmaris longus is absent in about 20 percent of Europeans, but is absent in only 3 percent of Orientals. Table 6 lists those muscles of the forearm that show pronounced differences among the continental groups of men in their dimensions. It is therefore highly probable that the position of rest of the hand will be quite different in individuals from these various populations. Even so, it may well be possible to accommodate these differences and so solve the problem of keyboard design by suitably positioning the hand relative to the keyboard.

EXAMPLE 2: SCHOOL FURNITURE SIZES

A second example of the solution of a problem of intrapopulation size variation is provided by school furniture. In the late 1940s and early 1950s

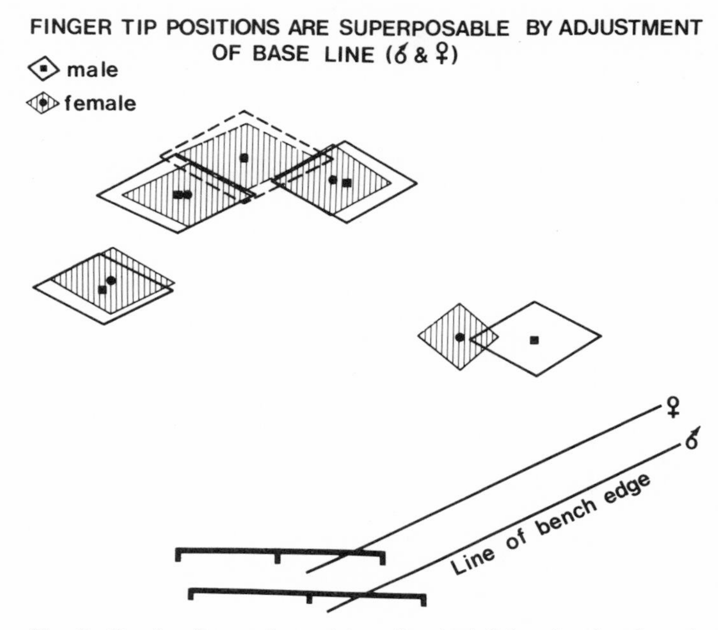

Fig. 5. Resting finger tip positions for the left hands of male and female subjects. The solid dots and squares are means, the points of the diamonds are one standard deviation from the mean.

the British Standards Institution (BSI) undertook a revision of standards for school furniture. To this end, basic information on the dimensions of body parts in children of different ages was derived from the Birmingham anthropometric survey and surveys elsewhere in Britain. The detailed information available on adult statures in the British population, and the rather less comprehensive information available on children's stature, enabled us to identify the major factors related to body size, for example, age, type of school, residential district, and region of the country. The major source of variation was, of course, age. Variations due to age were

Table 6. Muscles of the hand and forearm whose dimensions, proportions, and insertions vary significantly among races

Pronator teres	Brachioradialis
Flexor carpi radialis	Extensores carpi radialis longus et brevis
Palmaris longus	Extensor digitorum manus communis longus
Flexor carpi ulnaris	Extensor indicis proprius
Flexor digitorum sublimis	Extensor digiti V manus proprius
Flexor digitorum profundus	Extensor carpi ulnaris
Flexor pollicis longus	Abductor pollicis longus
Pronator quadratus	Extensor pollicis brevis

large in comparison with individual variation within age groups and gross in comparison with the effects of all other factors. Nonetheless, the cumulative effects of factors apart from age may be considerable, amounting in extreme instances to differences in mean stature for a particular age group of as much as 7.6 cm. In the smaller dimensions used in determining chair and table sizes, however, the cumulative effect of these other factors is likely to be absorbed in the broad tolerance required to provide access to the furniture and to accommodate individual variation.

Constructional dimensions for the furniture were derived from the basic anthropometric data and from clinical, anatomical, physiological, and related studies. These dimensions were evaluated in the light of manufacturing requirements, such as the costing advantages to be gained by having a minimum number of different sizes of spars. Based on such considerations, a decision was made to accommodate age variation by different sizes of furniture, and at the same time to use the different sizes of furniture to cover variations in size within age groups. Since the dimensions incorporated into the furniture related primarily to longitudinal segments of the body, it was possible to relate the sizes of furniture to the statures of the individuals for whom they would prove suitable. Thus percentages were calculated of each size of furniture needed to provide satisfactory accommodation for children in infant, junior, and senior schools. Frequencies of various sizes of furniture were also set out for a range of mean statures for each grade of school. Suitable percentages for yearly age groups were suggested, again linked with stature as a means for overcoming variation from school to school, to aid in the distribution of furniture within schools (B.S.I. 1959).

Here then is an illustration of how the great variation in size and shape that occurs during growth could be satisfactorily accommodated. The whole range of sizes and shapes of British children between the ages of five and sixteen years could be accommodated with five sizes of furniture. Such a rational system of sizing, based on adequate internationally coordinated dimensional and ergonomic surveys, may be a model that will enable us to satisfy the lesser variation that occurs among adult populations in different parts of the world.

CONCLUSION

Variations among human populations in size of the body and its parts are rarely greater than the variation that occurs within populations. Not only is interpopulation variation relatively limited, but an appreciable proportion of it follows a regular pattern, which may be of relevance to problems of design required for peoples in different parts of the world. Though there is little direct evidence on this point, such population differences appear to derive primarily from genetic differences.

The ergonomic implications of population variation vary according to the degree of complexity of the design problem and its sensitivity. In the more difficult problems there seems to be little alternative to the detailed investigation of the consumer population, and to designing specifically for it. Even here, however, it may be possible to solve some problems by a radical departure from traditional design patterns. In many less critical problems, it appears possible to solve the problems of interpopulation variation in the same way as has been successfully achieved with some problems of intrapopulation size variation. Two examples are given. The first relates to the design of keyboards, intricately related to the position of rest of the fingers. Here the great variation among individuals within a population could be accommodated by a slight variation in the positioning of the operator's arm relative to the keyboard and by a slight modification of key shape. A second example concerns school furniture, where it was possible to accommodate the great variation in size and shape that occurs in children aged five to sixteen in Britain by a limited number of sizes of furniture. The simple expedient of relating the size of furniture to the statures of the individuals for whom they would prove suitable not only accommodated variations in age but also accommodated variations within ages, as well as differences in size that characterize different types of school, socioeconomic class, and parts of the country.

These two solutions to intrapopulation differences—minor changes in positioning of the individual relative to the controls that he operates, and a rational system of sizing—are not the only solutions available. Adjustability of equipment, reliance on the ingenuity and adaptability of individuals, and a radical rethinking of complex problems are all possible ways of avoiding the economically unsatisfactory solution of individual design for each user population. But whatever the solution, it can only be achieved by efficient anthropometric and ergonomic surveys, so planned and coordinated that the very real differences among populations are not concealed by observational and technique differences. Here indeed there is room for international coordination and agreement.

REFERENCES

British Standards Institution. School furniture. Part 3: Pupils' classroom chairs and tables. British Standard 3030. London, 1959.

Buffon, G. L. *Oeuvres philosophiques: Histoire naturelle, générale et particulière*. Paris: Imprimerie Royale, 1766.

Gusinde, M. *Urwaldmenschen am Ituri*. Vienna: Springer-Verlag, 1948.

Kurth, G., and Lam, T. H. Körperwuchsformen und Arbeitsmittel; Angewandte Anthropometrie bei der Entwicklungshilfe. *Homo*. 1969, **20**, 209–221.

Le Gros Clark, W. E. *Fitting man to his environment*. Thirty-first Earl Gray Memorial Lecture. Durham, England: Durham University Press, 1949.

Ministry of Housing and Local Government. *Some aspects of designing for old people*. Design Bulletin No. 1. London: Her Majesty's Stationery Office, 1962.
Quetelet, A. *Anthropometrie*. Brussels: C. Muquardt, 1870.
Roberts, D. F. Basal metabolism, race and climate. *Journal of the Royal Anthropological Institute*, 1952, **82**, 169–83.
Roberts, D. F. Industrial applications of body measurements. *American Anthropologist*, 1956, **58**, 526–35.
Roberts, D. F. Neue Anwendungen der Anthropometrie. *Homo*, 1959, **10**, 40–46.
Roberts, D. F. Human—and inhuman—factors in office design. *Proceedings of the Royal Society of Health*, 1962, **62**, 78–83.
Roberts, D. F. Ergonomics and the industrial worker. *Occupational Health*, 1964, **16**, 17–22.
Roberts, D. F. *Climate and human variation*. Reading, Massachusetts: Addison-Wesley Publishing Co., 1972.
Shapiro, H. *Migration and environment*. London: Oxford University Press, 1939.
Tildesley, M. L. The relative usefulness of various characters on the living for racial comparison. *Man*, 1950, **50**, 14–18.
Wood Jones, F. *Anatomy of the hand*. London: Longman, 1920.

THREE

Anthropometric Measurements on Selected Populations of the World

ROBERT M. WHITE

A fundamental concept in the field of human factors engineering or ergonomics is the systems approach. According to this concept, the man together with his equipment, whether it be personal equipment he is wearing or using, or a machine he is operating, is considered to be a man-equipment system. A basic requirement for the efficient use and operation of such a system is that the man and the equipment be compatible. Since anthropometric data constitute a basic requisite for defining the elements of body size in the human engineering of man-equipment systems, anthropometry provides an essential input in the development of such systems.

SOURCES OF ANTHROPOMETRIC DATA

UNITED STATES

Most of the active research in anthropometry, at least in the United States, is carried out or sponsored by the Armed Forces. One reason for this is that the personnel of the Armed Forces represent an ideal, as well as unique, "population laboratory" in that the men are readily available for measurement. The resulting anthropometric data then can be applied in the development of military materiel.

The collection, analysis, and application of anthropometric data is not a recent development. Large numbers of U.S. Army troops were measured during and at the end of World War I, and the data were analyzed primarily

Robert M. White, Clothing and Personal Life Support Equipment Laboratory, U.S. Army Natick Laboratories, Natick, Massachusetts, U.S.A.

for clothing sizing (Davenport & Love 1921). A considerable amount of work in applied anthropometry was also carried on in the United States during and following World War II, particularly on aircraft cockpits, gun turrets, oxygen masks, and flight clothing (Randall et al. 1946). Only in recent years, however, has applied anthropometry become identified with human factors engineering in the United States.

In 1946 the U.S. Army Quartermaster Corps carried out an extensive anthropometric survey in which over 100,000 men were measured during their separation from the Army (Newman & White 1951), and in 1950 the U.S. Air Force conducted an anthropometric survey of about 4,000 flight personnel (Hertzberg et al. 1954). Between 1965 and 1970 new anthropometric surveys were carried out in which over 19,000 men of all the U.S. Armed Forces were measured. Both the report of the Army survey (White & Churchill 1971) and the one on Army aviators (Churchill et al. 1971) have been published; other reports are in preparation. Thus, during the past twenty-five years, a large amount of anthropometric data has been collected on U.S. military personnel.

Unfortunately, very few comparable data are available on the civilian population in the United States. A survey of some 15,000 women was conducted in 1939–40 by the Bureau of Home Economics of the U.S. Department of Agriculture, and the resulting data were used in the development of sizing and patterns for women's clothing (O'Brien & Shelton 1941). More recently, during the National Health Survey of 1960–62, carried out by the Public Health Service of the U.S. Department of Health, Education, and Welfare, 18 body measurements were taken on a carefully selected sample of about 3,000 men and 3,500 women from the U.S. civilian population (Stoudt et al. 1965, 1970).

COUNTRIES OTHER THAN THE UNITED STATES

Of greater importance and of considerably more interest for this book is the recent increase in the availability of anthropometric data from countries other than the United States. Anthropometry is hardly new in Europe, as it has been practiced in many countries for a long time. However, the results of large-scale anthropometric surveys, especially from other parts of the world, have become available only comparatively recently.

By large-scale surveys I mean surveys in which a large number of measurements is taken on large population samples. There are many references in the literature to studies in which a few specialized measurements or dimensions are taken on a small number of subjects. Such results are useful for specific purposes, but it is often difficult to generalize from limited data of this kind. Only fairly large anthropometric surveys provide the range or variety of body measurements on adequate numbers of people to generate the data necessary for many types of application in human engineering.

To be sure, some information concerning human variation in body size may be inferred from data on only height and weight. For example, a series of nutritional studies in which height and weight were recorded was carried out between 1956 and 1962 in some twenty-five countries in all parts of the world. The subjects utilized in these surveys were usually military personnel; however, in some countries a civilian sample also was studied. The results provide an interesting summary of human variation in height and weight, but the data are of limited value in terms of human engineering applications. The so-called ponderal index, calculated by dividing height by the cube root of weight, provides one general measure of body proportion. However, a great deal more information on variability in body size and proportions is necessary for satisfactory design work in the field of man-equipment systems.

In 1960–61 anthropometric surveys were conducted on military personnel in Turkey, Greece, and Italy, under the auspices of NATO's Advisory Group for Aeronautical Research and Development (AGARD). A report on these surveys by Hertzberg and his team of coworkers was published in book form by AGARD (Hertzberg et al. 1963). Surveys have been carried out on Republic of Korea Air Force pilots (Kay 1961) and Japanese Air Self-Defense Force pilots (Oshima et al. 1962). Large-scale anthropometric surveys have been conducted on the Royal Thai Armed Forces (White 1964*a*) and the Armed Forces of the Republic of Vietnam (White 1964*b*). Several surveys have been carried out with the Republic of Korea Armed Forces (Hart, Rowland, & Malina 1967; Kennedy & Rowland 1970). A survey to collect anthropometric data on Central and South American military personnel was initiated in 1965 at the U.S. Army Tropic Test Center in the Canal Zone (Dobbins & Kindick 1967). An anthropometric survey of the Imperial Iranian Armed Forces was carried out by the Iranian Army in 1968, with technical assistance from the United States (Noorani & Dillard 1970).

A large amount of anthropometric data is available from various European countries. For example, Air Force pilots have been measured in France (Ducros 1955) and in Belgium (Evrard 1954). In the Federal Republic of Germany, anthropometric studies have been carried out in the Army, including tanker personnel (Goltz & Platz 1965), and in the Air Force. Comparable anthropometric data are available on Norwegian men (Udjus 1964). In the United Kingdom, where research in anthropometry was carried on for many years by Morant, similar work is continuing, especially in the Royal Air Force at Farnborough (Samuel & Smith 1965; Simpson & Bolton 1968). Comparable anthropometric data also are available on Canadian pilots (Samuel & Smith 1965) and on Australian Army personnel (Army Inspection Service 1970).

The foregoing review of sources of anthropometric data is by no means exhaustive. These few references, however, illustrate the various populations that have been studied anthropometrically.

ANTHROPOMETRIC DATA FOR CERTAIN NATIONAL GROUPS

Body weight and stature are two basic measurements of the human body that reflect both volume and height. Sitting height is important in the design of vehicles, aircraft, control consoles, and other types of equipment. Chest circumference is a basic girth measurement of the body, widely used in the sizing of clothing. Statistics on these four body measurements are presented in Tables 1 through 8 for a number of populations.[1]

Tables 1 through 4 show the sample, the date, the number of men measured (*N*), the mean age of the men (in years), and the mean and standard deviation for the anthropometric measurement. In these tables, the samples have been arranged or ranked in increasing order of their means, from smallest to largest.

Tables 5 through 8 show certain percentile values for some of the same samples as in Tables 1 through 4. Again, the samples have been ranked in increasing order of the 50th percentile or median.

In comparing variations in body size, consideration should be given to the ages of the samples, since body size is, to some extent at least, correlated with age. Although mean ages are not available for all the samples, most of them are between twenty and thirty years for the groups cited here.

RANGES OF VARIATION

The mean weights (Table 1) vary from 51.1 kg (112.7 lbs) for the Vietnamese to 78.7 kg (173.6 lbs) for U.S. Air Force flyers, representing a spread of 27.6 kg (60.9 lbs).

Mean statures (Table 2) also show considerable variation. The lowest mean stature is that for the Vietnamese (160.5 cm or 63.2 inches), while the highest is for Belgian flight personnel (179.9 cm or 70.8 inches). Thus, the range of means for stature is 19.4 cm (7.6 inches).

Mean values for sitting height (Table 3) vary from 85.0 cm (33.5 inches) to 93.2 cm (36.7 inches), representing a range of 8.2 cm (3.2 inches). Sitting height also illustrates one interesting variation in human body proportions. A large part of the variation in human stature is in the length of the legs; the torso is relatively constant in length. If sitting height is expressed as a percentage of stature, the resulting ratio or index is about 52 percent for Americans and most Europeans. Thus leg length is approximately 48 percent of stature. However, in the case of the Korean and Japanese pilots cited here, sitting height is over 54 percent of stature, so that leg length is only about 46 percent of stature.[2]

[1] *Editor's note*: Hertzberg (1972) summarizes some of the same statistics as are given in these tables and gives some limited statistics on populations other than those described here. When discrepancies occur between his data and those here, White believes the former to be in error.

[2] *Editor's note*: See also the papers by Kennedy (pages 53–54) and by Roberts (page 12) on this point.

Table 1. Statistics on the weights (in kg) and certain other characteristics of 26 samples

Sample	Date	N	Age[a]	Weight Mean	Weight SD
Republic of Vietnam Armed Forces	1964	2,128	27.2	51.1	6.0
Thailand Armed Forces	1964	2,950	24.0	56.3	5.8
India Army	1969	4,000	27.0	57.2	5.7
Republic of Korea Army	1970	3,473	24.7	60.3	5.1
Japan JASDF pilots	1962	239	24.1	61.1	5.9
Iran Armed Forces	1970	9,414	23.8	61.6	7.7
Republic of Korea ROKAF pilots	1961	264	28.0	62.7	6.5
Latin America Armed Forces (18 countries)	1967	733	23.1	63.4	7.7
Turkey Armed Forces	1963	915	24.1	64.6	8.2
U.S. Army WWI demobilization	1921	79,706	24.9	65.6	7.7
France Flight personnel	1955	1,000	18-45	65.8	7.0
Greece Armed Forces	1963	1,084	22.9	67.0	7.6
Australia Army	1970	3,695	21.0	68.5	8.4
Belgium Flight personnel	1954	2,214	17-50	68.6	7.8
Norway Young Men	1964	5,765	20.0	70.1	7.5
U.S. Army WWII separatees	1951	24,506	24.3	70.2	9.3
Italy Armed Forces	1963	1,358	26.5	70.3	8.4
United Kingdom RAF pilots	1965	4,357	—	71.9	9.2
U.S. Army Ground troops	1971	6,677	22.2	72.2	10.6
Fed. Rep. of Germany Army tank crews	1965	298	22.8	72.3	8.1
U.S. Air Force Flight personnel	1954	4,052	27.9	74.2	9.5
United Kingdom RAF and RN air crew	1968	200	28.7	74.3	9.4
U.S. civilian men Nat'l. Health Survey	1965	3,091	44.0	75.3	12.6
Canada RCAF pilots	1965	314	—	76.4	9.9
U.S. Army Aviators	1971	1,482	26.2	77.6	10.8
U.S. Air Force Flight personnel	1972	2,420	30.0	78.7	9.7

[a]Mean values except where ranges are given.

Table 2. Statistics on the statures (in cm) and certain other characteristics of 26 samples

Sample	Date	N	Age[a]	Stature Mean	Stature SD
Republic of Vietnam Armed Forces	1964	2,129	27.2	160.5	5.5
Thailand Armed Forces	1964	2,950	24.0	163.4	5.3
Republic of Korea Army	1970	3,473	24.7	164.0	5.9
Latin America Armed Forces (18 countries)	1967	733	23.1	166.4	6.1
Iran Armed Forces	1970	9,414	23.8	166.8	5.8
Japan JASDF pilots	1962	239	24.1	166.9	4.8
India Army	1969	4,000	27.0	167.5	6.0
Republic of Korea ROKAF pilots	1961	264	28.0	168.7	4.6
Turkey Armed Forces	1963	915	24.1	169.3	5.7
Greece Armed Forces	1963	1,084	22.9	170.5	5.9
Italy Armed Forces	1963	1,358	26.5	170.6	6.2
France Flight personnel	1955	7,084	18-45	171.3	5.8
U.S. Army WWI demobilization	1921	96,596	24.9	172.0	6.7
Australia Army	1970	3,695	21.0	173.0	6.0
U.S. civilian men Nat'l. Health Survey	1965	3,091	44.0	173.2	7.2
U.S. Army WWII separatees	1951	24,508	24.3	173.9	6.4
U.S. Army Ground troops	1971	6,682	22.2	174.5	6.6
U.S. Army Aviators	1971	1,482	26.2	174.6	6.3
Fed. Rep. of Germany Army tank crews	1965	300	22.8	174.9	6.1
U.S. Air Force Flight personnel	1954	4,062	27.9	175.5	6.2
United Kingdom RAF and RN air crew	1968	200	28.7	177.0	6.1
United Kingdom RAF pilots	1965	4,357	—	177.2	6.2
U.S. Air Force Flight personnel	1972	2,420	30.0	177.3	6.2
Canada RCAF pilots	1965	314	—	177.4	6.1
Norway Young men	1964	5,765	20.0	177.5	6.0
Belgium Flight personnel	1954	2,450	17-50	179.9	5.8

[a] Mean values except where ranges are given.

Table 3. Statistics on the sitting heights (in cm) and certain other characteristics of 21 samples

Sample	Date	N	Mean Age	Sitting Height Mean	Sitting Height SD
Republic of Vietnam Armed Forces	1964	2,124	27.2	85.0	3.3
Thailand Armed Forces	1964	2,950	24.0	86.4	3.1
Latin America Armed Forces (18 countries)	1967	733	23.1	86.9	—
Iran Armed Forces	1970	9,414	23.8	87.9	3.3
Republic of Korea Army	1970	3,473	24.7	88.7	3.6
Italy Armed Forces	1963	1,358	26.5	89.7	3.2
Turkey Armed Forces	1963	915	24.1	89.7	3.2
Greece Armed Forces	1963	1,084	22.9	90.2	3.0
U.S. Army WWI demobilization	1921	96,239	24.9	90.4	3.5
U.S. civilian men Nat'l. Health Survey	1965	3,091	44.0	90.4	3.8
U.S. Army Ground troops	1971	6,682	22.2	90.7	3.7
Japan JASDF pilots	1962	239	24.1	90.8	2.6
Republic of Korea ROKAF pilots	1961	264	28.0	90.8	2.8
U.S. Army WWII separatees	1951	24,352	24.3	90.9	3.4
U.S. Army Aviators	1971	1,482	26.2	90.9	3.2
U.S. Air Force Flight personnel	1954	4,061	27.9	91.3	3.3
Fed. Rep. of Germany Army tank crews	1965	281	22.8	91.3	3.4
Norway Young men	1964	5,765	20.0	92.6	3.5
United Kingdom RAF and RN air crew	1968	200	28.7	92.7	3.0
Canada RCAF pilots	1965	314	—	93.2	3.0
U.S. Air Force Flight personnel	1972	2,420	30.0	93.2	3.2

Table 4. Statistics on the chest circumferences (in cm) and certain characteristics of 23 samples

Sample	Date	N	Age[a]	Chest circumference	
				Mean	SD
Republic of Vietnam Armed Forces	1964	2,127	27.2	81.1	4.3
Thailand Armed Forces	1964	2,950	24.0	85.0	4.1
Japan JASDF pilots	1962	239	24.1	88.0	3.9
Republic of Korea Army	1970	3,473	24.7	88.3	4.8
U.S. Army WWI demobilization	1921	95,867	24.9	88.8	5.1
India Army	1969	4,000	27.0	90.0	4.4
Iran Armed Forces	1970	9,414	23.8	90.4	5.5
France Flight personnel	1955	1,000	18-45	90.9	5.2
Republic of Korea ROKAF pilots	1961	264	28.0	91.0	5.4
Belgium Flight personnel	1954	2,450	17-50	91.0	—
Turkey Armed Forces	1963	915	24.1	91.1	5.0
U.S. Army WWII separatees	1951	24,470	24.3	92.4	6.0
Greece Armed Forces	1963	1,084	22.9	92.5	5.1
U.S. Army Ground troops	1971	6,682	22.2	93.8	6.7
Australia Army	1970	3,695	21.0	94.5	5.9
Fed. Rep. of Germany Army tank crews	1965	300	22.8	94.8	—
Italy Armed Forces	1963	1,358	26.5	95.0	5.3
Canada RCAF pilots	1965	314	—	95.2	6.2
United Kingdom RAF and RN air crew	1968	200	28.7	97.9	5.8
U.S. Army Aviators	1971	1,482	26.2	98.4	6.9
U.S. Air Force Flight personnel	1954	4,057	27.9	98.6	6.2
U.S. Air Force Flight personnel	1972	2,420	30.0	98.6	6.4
U.S. civilian men Nat'l. Health Survey	1970	3,091	44.0	99.6	8.2

[a] Mean values except where ranges are given.

Sample	Date	Percentiles										
		1st	2nd	5th	10th	25th	50th	75th	90th	95th	98th	99th
Republic of Vietnam Armed Forces	1964	39.8	40.5	42.4	44.2	46.7	50.6	54.7	58.5	61.5	66.5	70.0
Thailand Armed Forces	1964	45.5	46.5	48.0	50.0	52.5	56.0	60.0	64.0	67.0	70.5	73.0
Republic of Korea Army	1970	47.6	48.9	51.2	53.0	55.7	59.8	63.4	67.5	69.8	72.5	74.3
Japan JASFD pilots	1962	50.3	50.9	52.5	54.2	56.8	60.6	64.6	69.0	71.8	75.2	79.9
Iran Armed Forces	1970	46.8	48.4	50.8	52.9	56.5	60.8	65.8	71.4	75.6	81.4	85.9
Republic of Korea ROKAF pilots	1961	50.1	51.6	53.1	54.7	58.4	62.1	66.6	72.1	76.6	80.6	82.6
Latin America Armed Forces (18 countries)	1967	47.6	—	52.2	—	58.1	63.0	68.0	—	78.0	—	85.7
Turkey Armed Forces	1963	48.2	49.9	52.5	54.8	58.9	63.8	69.5	75.3	79.2	84.1	87.6
Greece Armed Forces	1963	52.2	53.6	55.8	57.8	61.6	66.4	71.8	77.1	80.5	84.2	86.7
Australia Army	1970	53.5	—	56.5	59.0	63.0	68.5	75.0	80.5	84.0	—	88.0
U.S. Army WWII separatees	1951	51.7	53.5	56.2	59.0	64.0	69.4	75.8	82.1	87.1	93.0	97.5
Italy Armed Forces	1963	53.9	55.2	57.6	60.0	64.3	69.6	75.4	81.2	85.0	89.7	93.1
U.S. Army Ground troops	1971	52.6	54.5	57.4	60.0	64.8	71.0	78.4	86.3	91.6	98.3	103.0
United Kingdom RAF and RN air crew	1968	58.1	59.4	60.6	62.1	66.9	73.0	80.7	86.6	88.9	94.6	97.5
U.S. Air Force Flight personnel	1954	55.8	57.2	60.1	62.7	67.4	73.4	80.1	87.4	91.1	96.0	97.9
U.S. civilian men Nat'l. Health Survey	1965	49.9	52.2	56.2	59.9	66.7	74.4	83.0	92.1	97.5	103.9	108.4
U.S. Army Aviators	1971	55.5	57.2	60.4	63.7	69.9	77.4	84.9	91.8	96.0	101.0	104.4
U.S. Air Force Flight personnel	1972	57.9	60.2	63.6	66.6	71.9	78.2	85.0	91.5	95.6	100.2	103.3

Table 6. Certain percentiles for stature (in cm) of 18 of the samples in Table 2

Sample	Date	Percentiles										
		1st	2nd	5th	10th	25th	50th	75th	90th	95th	98th	99th
Republic of Vietnam Armed Forces	1964	148.1	149.5	151.6	153.4	156.8	160.4	164.3	167.7	169.6	171.5	173.0
Thailand Armed Forces	1964	151.5	153.0	155.0	157.0	160.0	163.5	167.0	170.0	172.0	174.0	176.0
Republic of Korea Army	1970	150.2	151.9	154.5	156.9	160.2	164.0	167.6	170.8	172.7	174.3	176.2
Republic of Korea ROKAF pilots	1961	157.6	159.5	159.6	161.6	163.7	165.6	171.6	173.6	173.7	177.2	177.9
Latin America Armed Forces (18 countries)	1967	153.7	—	157.0	—	162.1	165.9	170.1	—	177.0	—	181.6
Iran Armed Forces	1970	153.5	155.1	157.5	159.6	162.9	166.7	170.7	174.4	176.6	179.1	180.7
Japan JASDF pilots	1962	157.3	157.9	159.4	161.0	163.6	166.7	169.7	173.0	175.0	177.8	180.4
Turkey Armed Forces	1963	157.8	158.9	160.6	162.2	165.2	169.0	173.2	177.1	179.2	181.3	182.4
Greece Armed Forces	1963	157.5	158.8	160.9	162.9	166.5	170.5	174.4	178.0	180.3	182.9	184.8
Italy Armed Forces	1963	157.0	158.1	160.2	162.3	166.3	170.7	174.8	178.5	180.7	183.5	185.6
Australia Army	1970	161.0	—	164.0	166.0	169.0	173.0	178.0	181.5	184.0	—	188.0
U.S. civilian men Nat'l. Health Survey	1965	156.7	158.8	161.5	163.8	168.7	173.5	178.1	182.4	184.9	188.0	189.5
U.S. Army WWII separatees	1951	159.3	160.8	163.3	165.6	169.7	174.0	178.3	182.4	184.4	187.2	189.2
U.S. Army Ground troops	1971	158.9	160.9	163.8	166.2	170.1	174.4	178.9	183.0	185.6	188.4	190.3
U.S. Army Aviators	1971	160.5	161.8	164.2	166.4	170.3	174.6	178.8	182.6	185.0	188.0	190.2
U.S. Air Force Flight personnel	1954	161.3	163.0	165.5	167.6	171.4	175.6	179.7	183.5	185.8	188.5	190.3
United Kingdom RAF and RN air crew	1968	163.1	164.6	166.1	168.4	172.7	177.0	181.1	184.4	185.9	190.0	192.8
U.S. Air Force Flight personnel	1972	163.2	164.8	167.2	169.4	173.1	177.3	181.5	185.4	187.7	190.2	191.8

Table 7. Certain percentiles for sitting height (in cm) of 17 of the samples in Table 3

Sample	Date	Percentiles										
		1st	2nd	5th	10th	25th	50th	75th	90th	95th	98th	99th
Republic of Vietnam Armed Forces	1964	77.5	78.5	79.6	80.9	82.8	85.0	87.3	89.3	90.5	91.5	92.5
Thailand Armed Forces	1964	79.5	80.3	81.5	82.5	84.5	86.5	88.5	90.5	91.5	93.0	93.5
Latin America Armed Forces (18 countries)	1967	78.4	—	80.9	—	84.7	86.9	88.9	—	91.9	—	94.6
Iran Armed Forces	1970	80.3	81.2	82.5	83.6	85.6	87.8	90.1	92.1	93.3	94.7	95.6
Republic of Korea Army	1970	82.5	83.1	83.9	85.1	87.0	88.8	90.8	92.4	93.4	94.5	95.1
Turkey Armed Forces	1963	83.1	83.8	84.8	85.8	87.5	89.6	91.8	93.9	95.1	96.4	97.3
Italy Armed Forces	1963	82.0	82.9	84.2	85.5	87.5	89.7	91.8	93.7	94.8	96.1	97.1
Greece Armed Forces	1963	83.3	84.1	85.4	86.4	88.2	90.2	92.3	94.2	95.2	96.3	97.0
Japan JASDF pilots	1962	84.5	85.1	86.2	88.0	89.1	90.7	92.4	94.0	95.3	96.6	97.5
Republic of Korea ROKAF pilots	1961	83.3	84.6	86.1	87.0	88.7	90.7	93.0	94.7	95.3	97.1	98.2
U.S. civilian men Nat'l. Health Survey	1965	81.0	82.3	84.3	85.9	88.1	90.7	93.2	95.5	96.5	98.0	98.8
U.S. Army Ground troops	1971	82.0	83.0	84.5	85.9	88.2	90.8	93.2	95.4	96.7	98.2	99.2
U.S. Army Aviators	1971	83.3	84.3	85.7	86.8	88.8	90.9	93.1	95.1	96.3	97.7	98.6
U.S. Army WWII separatees	1951	82.8	83.8	85.1	86.4	88.4	90.9	93.2	95.2	96.5	98.0	99.1
U.S. Air Force Flight personnel	1954	83.5	84.6	85.8	87.2	89.1	91.4	93.6	95.5	96.6	98.0	98.9
United Kingdom RAF and RN air crew	1968	86.1	86.9	87.6	88.4	90.4	92.5	95.0	96.8	97.5	98.6	98.8
U.S. Air Force Flight personnel	1972	86.2	87.0	88.1	89.2	91.0	93.1	95.3	97.4	98.6	99.8	100.6

Table 8. Certain percentiles for chest circumference (in cm) of 17 of the samples in Table 4

Sample	Date	Percentiles										
		1st	2nd	5th	10th	25th	50th	75th	90th	95th	98th	99th
Republic of Vietnam Armed Forces	1964	72.2	73.0	74.5	75.9	78.2	80.8	83.7	86.7	88.5	91.0	93.2
Thailand Armed Forces	1964	76.0	77.0	78.5	80.0	82.0	85.0	87.5	90.0	92.0	94.0	96.0
Japan JASDF pilots	1962	80.1	80.6	82.1	83.1	85.1	87.8	90.6	93.0	95.0	97.4	99.2
Republic of Korea Army	1970	78.2	79.9	81.2	82.9	85.4	88.0	91.0	94.2	96.0	98.0	99.0
Iran Armed Forces	1970	79.2	80.5	82.4	84.0	86.7	89.9	93.4	97.4	100.2	104.1	107.1
Republic of Korea ROKAF pilots	1961	81.1	81.9	83.8	84.8	87.3	90.3	93.5	97.6	101.4	106.7	108.0
Turkey Armed Forces	1963	80.6	81.9	83.7	85.2	87.7	90.7	94.2	97.8	100.1	103.0	105.0
U.S. Army WWII separatees	1951	80.3	81.5	83.6	85.3	88.4	91.9	96.0	100.1	102.9	106.4	109.2
Greece Armed Forces	1963	81.2	82.7	84.7	86.4	89.0	92.2	95.7	99.2	101.5	104.1	105.9
U.S. Army Ground troops	1971	80.9	82.2	84.1	85.9	89.1	93.0	97.7	102.6	105.9	109.9	112.8
Australia Army	1970	83.5	—	86.0	88.0	91.0	94.5	98.5	103.0	105.5	—	109.0
Italy Armed Forces	1963	84.9	85.7	87.1	88.6	91.3	94.6	98.0	101.7	104.3	107.8	110.5
United Kingdom RAF and RN air crew	1968	86.4	87.6	89.4	90.9	93.7	97.0	101.6	105.9	108.0	111.8	112.8
U.S. Air Force Flight personnel	1954	85.7	87.0	89.1	91.1	94.2	98.2	102.5	106.9	109.7	112.2	113.9
U.S. Army Aviators	1971	84.5	85.5	87.5	89.6	93.6	98.2	102.9	107.2	109.9	113.3	115.9
U.S. Air Force Flight personnel	1972	84.8	86.3	88.6	90.6	94.1	98.3	102.7	106.8	109.4	112.4	114.4
U.S. civilian men Nat'l. Health Survey	1970	82.6	84.3	87.1	89.4	94.0	99.1	104.6	110.5	114.0	118.4	121.2

Chest circumference is a primary girth measurement of the body and is highly correlated with body weight. Mean chest circumferences (Table 4) vary from 81.1 cm (31.9 inches) for the Vietnamese to 99.6 cm (39.2 inches) for U.S. civilian men. As another measure of body proportion, mean chest circumference ranges from about 50 percent to over 57 percent of stature for different samples.

PERCENTILES

The anthropometric data presented in Tables 1–4 are interesting from the viewpoint of the overall range of body dimensions. The arithmetic mean is perhaps the most commonly used and generally recognized average. While the mean is relatively simple to compute, it has the disadvantage of being distorted by extreme values. Although the mean is a convenient indication of central tendency, it is virtually useless in terms of practical application. Percentile values for body dimensions are of greater interest and more importance in terms of practical applications. They are useful in depicting the spread or range of a dimension and may be used to estimate the degree of coverage or the range of adjustability required for a given population.

Percentile values for weight are shown in Table 5. The lowest 1st percentile value is for the Vietnamese (39.8 kg or 87.8 lbs), while the highest 1st percentile is for British RAF and RN air crewmen (58.1 kg or 128.0 lbs). The lowest 99th percentile is 70.0 kg (154.4 lbs), again for Vietnamese, and the highest 99th percentile is 108.4 kg (239.0 lbs) for U.S. civilian men.

Similar comparisons of percentile values may be made for stature (Table 6), sitting height (Table 7), and chest circumference (Table 8).

APPLICATION OF ANTHROPOMETRIC DATA IN HUMAN ENGINEERING

That variability in human body size exists is, of course, well-known. The basic question is how this problem should be met intelligently and efficiently by those concerned with the man-equipment interface.

First, accurate and reliable anthropometric data should be available on the population for which equipment is intended. Second, the data must be used and applied carefully. At this point the state-of-the-art cannot claim ready answers and standard solutions for the great variety of problems that arise in human engineering. Variability in human body size is not a new problem, since it has been anticipated and discussed previously by others. But it should receive renewed emphasis. Two examples from the literature are worth mentioning here.

In 1968 Hertzberg wrote an article entitled: "World Diversity in Human Body Size and Its Meaning in American Aid Programs" (Hertzberg 1968). In that paper he compared percentile values for several body dimensions on

Italian, Greek, Turkish, Korean, Japanese, and U.S. Air Force personnel. He concluded that:

> Not only is the large majority of the flying personnel among our Mediterranean and Oriental allies below our 50th percentile in most dimensions, showing that their sizes are different from ours, but also their proportions are different. Hence, smaller sizes of garments which fit us will not necessarily fit all of them adequately. In any case, it is inescapable that reliable anthropometric data on such populations are essential to the effective solution of the vexing and expensive fitting problems associated with clothing and equipment furnished our allies under our aid programs.

Although his statement has to do with a highly specialized application in military flight clothing and equipment, similar comments may be made concerning many other types of human factors engineering.

Another example of the concern regarding variability in body size may be found in a paper entitled: "The Range of Anthropometric Measurements for Asian Populations," published in the United States by the Society of Automotive Engineers (1968). In this report, sponsored by the Construction and Industrial Machinery Technical Committee of the SAE, anthropometric data from Vietnam and Turkey were presented and discussed. The paper states that: "The employment of body dimensions as an integral part of equipment design requires data on the intended user population. This specific information is necessary because a large difference in body size may exist among groups of people."

With reference to construction and industrial machinery, the SAE paper states that:

> This report is intended to help eliminate many design-related costs by providing relevant Asian body dimension data at an early stage of equipment development. This information is intended to serve as a guideline for the design of efficiently operating equipment and is not meant to be an exhaustive checklist of operator-related design considerations.
>
> Body dimensions greatly influence the design of seats, the positioning of controls, and the dimensions and position of various vehicle control area features (for example, windshields and panels). The use of body dimensions as a design guideline will help provide an effective work place and a much safer, more comfortable, and healthy operator environment.
>
> If body dimensions and information concerning operator limitations and capabilities are integrated with operational requirements during early conceptualization and design stages, significant improvements in system operation can be achieved with minimum cost. Errors concerning the design of certain vehicle features which affect operator effectiveness will be avoided. Redesign costs will be reduced.

SUMMARY

The problem of variability in human body size has been discussed here in terms of its impact on human factors engineering. While the uses and applications of anthropometric data still involve many unanswered questions and unsolved problems, the fund of available data is, fortunately, growing. More useful data are now being accumulated on diverse populations. These data will be useful for improved and more efficient human engineering in man-equipment systems.

REFERENCES

Army Inspection Service, Headquarters. *Anthropometric survey; body dimensions—1970*. Melbourne, Australia: Australian Army, 1970.

Churchill, E., McConville, J. T., Laubach, L. L., and White, R. M. Anthropometry of U.S. Army aviators—1970. Technical Report 72-52-CE. Natick, Massachusetts: U.S. Army Natick Laboratories, 1971.

Davenport, C. B., and Love, A. G. Army Anthropology. In Vol. 15, Part 1, of *The Medical Department of the United States Army in the World War*. Washington: Office of the Surgeon General, Department of the Army, 1921.

Dobbins, D. A., and Kindick, C. M. Anthropometry of the Latin-American Armed Forces. Research Report No. 10 (Interim Report). Fort Clayton, Canal Zone: U.S. Army Tropic Test Center, 1967.

Ducros, E. Statistiques de biométrie médicale élémentaire relatives au personnel navigant de l'armée de l'air française. In *Anthropometry and human engineering*, AGARDograph No. 5. London: Butterworths Scientific Publications, 1955.

Evrard, E. *Statistiques biométriques relatives au personnel navigant de la force aérienne belge*. Belgium: Direction du Service de Santé, Force Aérienne, Ministère de la Défense Nationale, May 1954.

Goltz, E., and Platz, B. *Anthropometrische Studien*. Report T-418. Augsburg, Germany: Der Deutschen Entwicklungsgesellschaft, 1965.

Hart, G. L., Rowland, G. E., and Malina, R. Anthropometric survey of the Armed Forces of the Republic of Korea. Technical Report 68-8-PR. Natick, Massachusetts: U.S. Army Natick Laboratories, 1967.

Hertzberg, H. T. E. World diversity in human body size and its meaning in American aid programs. Arlington, Virginia: Office of Aerospace Research, *OAR Research Review*, 1968, **7** (12), 14–17.

Hertzberg, H. T. E. Engineering Anthropology. In Van Cott, H. P., and Kincade, R. G. (Eds.) *Human engineering guide to equipment design* (revised edition). Washington, D.C.: U.S. Government Printing Office, 1972.

Hertzberg, H. T. E., Churchill, E., Dupertuis, C. W., White, R. M., and Damon, A. *Anthropometric survey of Turkey, Greece, and Italy*. AGARDograph 73. New York: Pergamon Press, 1963.

Hertzberg, H. T. E., Daniels, G. S., and Churchill, E. Anthropometry of flying personnel—1950. WADC Technical Report 52-321. Wright-Patterson Air Force Base, Ohio: Aero Medical Laboratory, 1954.

Kay, W. C. Anthropometry of the ROKAF pilots. *Republic of Korea Air Force Journal of Aviation Medicine*, 1961, **9**, 61–113.

Kennedy, J. C., and Rowland, G. E. Analysis of the impact of body size upon Korean soldier performance with U.S. weapons and equipment. Report R&C-70-7-104. Haddonfield, New Jersey: Rowland and Company, 1970.

Newman, R. W., and White, R. M. Reference anthropometry of Army men. Report No. 180. Lawrence, Massachusetts: Environmental Protection Branch, U.S. Army Quartermaster Climatic Research Laboratory, 1951.

Noorani, S., and Dillard, C. N., Jr. Anthropometric survey of the Imperial Iranian Armed Forces. Technical Report of the Combat Research and Evaluation Center. Teheran, Iran: Imperial Iranian Ground Forces, 1970.

O'Brien, R., and Shelton, W. C. *Women's measurements for garment and pattern construction.* Bureau of Home Economics, Textiles and Clothing Division, Miscellaneous Publication No. 454. Washington, D.C.: U.S. Department of Agriculture and Work Projects Administration, December 1941.

Oshima, M., Fujimoto, T., Oguro, T., Tobimatsu, N., Mori, T., Tanaka, I., Watanabe, T., and Alexander, M. Anthropometry of Japanese pilots. *Reports of the Aero Medical Laboratory* (Tokyo, Japan), 1962, **2**, 70–114.

Randall, F. E., Damon, A., Benton, R. S., and Patt, D. I. Human body size in military aircraft and personal equipment. Army Air Forces Technical Report 5501. Wright Field, Dayton, Ohio: Air Materiel Command, June 10, 1946.

Samuel, G. D., and Smith, E. M. B. A comparison of seven anthropometric variables of American, British, and Canadian pilots. Report No. 322. Farnborough, Hants.: Royal Air Force Institute of Aviation Medicine, 1965.

Simpson, R. E., and Bolton, C. B. An anthropometric survey of 200 R.A.F. and R.N. aircrew and the application of the data to garment size rolls. Technical Report 67125. Farnborough, Hants.: Royal Aircraft Establishment, 1968.

Society of Automotive Engineers. *The range of anthropometric measurements for Asian populations.* SAE J317. New York, New York: Society of Automotive Engineers, 1968.

Stoudt, H. W., Damon, A., McFarland, R. A., and Roberts, J. *Weight, height, and selected body dimensions of adults—United States, 1960–1962.* Public Health Service Publication No. 1000—Series 11—No. 8. Washington, D.C.: U.S. Department of Health, Education, and Welfare, 1965.

Stoudt, H. W., Damon, A., McFarland, R. A., and Roberts, J. *Skinfolds, body girths, biacromial diameter, and selected anthropometric indices of adults—United States, 1960–1962.* Public Health Service Publication No. 1000—Series 11—No. 35. Washington, D.C.: U.S. Department of Health, Education, and Welfare, 1970.

Udjus, L. G. *Anthropometrical changes in Norwegian men in the twentieth century.* Norwegian Monographs on Medical Science. Oslo, Norway: Universitetsforlaget, 1964.

White, R. M. Anthropometric survey of the Royal Thai Armed Forces. Natick, Massachusetts: U.S. Army Natick Laboratories, 1964*a*.

White, R. M. Anthropometric survey of the Armed Forces of the Republic of Vietnam. Natick, Massachusetts: U.S. Army Natick Laboratories, 1964*b*.

White, R. M. The utilization of military anthropometry for aircraft cockpit design. In *Problems of the cockpit environment*, AGARD Conference Proceedings No. 55, NATO-AGARD, 1970.

White, R. M., and Churchill, E. The body size of soldiers; U.S. Army anthropometry—1966. Technical Report 72-51-CE. Natick, Massachusetts: U.S. Army Natick Laboratories, 1971.

FOUR

International Anthropometric Variability and Its Effects on Aircraft Cockpit Design

KENNETH W. KENNEDY

The anthropometric characteristics of consumer populations play an important role in the design of many products. A myriad of items from space capsules to earth movers, bathrooms to milling machines, and theodolites to submarines require variable levels of accommodation to the sizes and proportions of the human body. Great effort is generally devoted to this end, particularly when the product is to be sold on a competitive market. If the product is rather simple the effort is usually relatively inexpensive and successful. On the other hand, if the product is a highly complex system, such as an aircraft, the attainment of a high level of accommodation almost invariably requires expensive economic and engineering trade-offs with varying levels of success.

This paper is concerned with high performance, single seat, military aircraft cockpits and the problems encountered in accommodating them to the anthropometric requirements of foreign military users. These problems often are very difficult. Design changes invariably required to cope with any significant anthropometric differences are fraught with seemingly insurmountable economic and engineering problems. Still, malaccommodation in aircraft not only produces a condition in which the product is inconvenient to operate, but one in which the user's safety and the basic mission of the aircraft can be compromised.

Kenneth W. Kennedy, Aerospace Medical Research Laboratory, Wright-Patterson Air Force Base, Ohio, U.S.A.

THE SERIOUSNESS OF THE PROBLEM IN AIRCRAFT

At times it may be difficult to imagine the seriousness of the problems encountered when a cockpit is not designed specifically for the using population. One can brush these problems aside in the name of patriotism, expedience, or whatever, but in the final analysis missions fail and aircraft and air crews may be lost as a direct consequence of a relatively "inconsequential" or "unimportant" anthropometric design consideration.

The following is a fictional account composed of factual ingredients. It illustrates how the interplay of apparently minor inadequacies of design can and does lead to tragic accidents.

Imagine a pilot on a flying mission of 9 to 10 hours. Such missions are not uncommon. After a few hours in the air the pilot wishes he could stand and walk around a little to relieve the pressure on his backside and legs. He squirms in his seat, first sitting on one buttock, then the other. After a while he tries to push himself forward in the seat but is held back by the lap belt. He lowers, then raises the seat to change the angulation of his knees and to relieve the ache beginning to bother him under one of his thighs. In a little while the pain increases and spreads throughout both thighs. After another 2 hours it begins to become unbearable. He is counting the minutes to touch-down, wondering if he can bear the discomfort that long. Three more hours yet to go. The pain has now traveled to the small of his back. The small changes in body position possible in the seat no longer help relieve the pain.

When he finally makes his approach, one of his two engines dies. To maintain his course requires the immediate application of full left rudder and left aileron. Because his legs are so painful, he cannot maintain the necessary rudder pedal force, and his extended left leg prevents him from obtaining full left aileron.

He crashes off the side of the runway. He and his crew are severely injured.

Discomfort in this aircraft was viewed in the usual manner, that is, as relatively unimportant. In this instance, however, the pilot's discomfort was protracted over a long period of time and became so unbearable that it consumed his thoughts. The pain was eventually replaced by numbness. When it became necessary to apply a relatively large but manageable force to the rudder pedal to maintain the course of the aircraft, the required strength simply was not available.

Eighty-two kg (180 lbs) of force are required on this particular aircraft to depress a rudder pedal after the failure of one engine. Such a force is not too great, even for the smaller American and European pilots, under normal circumstances. In the situation just described, however, circumstances were far from normal. In addition to applying full rudder to compensate for the engine failure, it was also necessary to apply full aileron in the same direction. When a pilot's leg is fully extended to obtain full rudder, full aileron is unavailable to a substantial number of pilots for two reasons:

1) Control wheel movement in both directions is interfered with by the *normal* positioning of the pilot's thighs. The interference is compounded when the pilot extends his leg to apply full rudder.

2) The control column is displaced to the left of the centerline of the pilot's seat, thereby increasing the restriction to aileron control on that side.

At least two general classes of problem were responsible for the unfortunate results in this account: comfort and accommodation to the anthropometry of the pilot. However, these are separate problems only in theory. In practice they merge at so many points, insofar as the aircraft cockpit is concerned, that they must be considered together. Comfort cannot be achieved without the application of sound engineering anthropometric design techniques. Even so, providing an optimum dimensional relationship between the pilot and the structures and equipment within a cockpit does not assure comfort. In the case of very long flights, ordinary standards of comfort are not good enough. Flights of long duration require the development of new devices to provide for extra margins of comfort far beyond that which is adequate for short flights. Here, however, I shall not be concerned with comfort per se, but with the anthropometric considerations essential to the basic geometric layout of cockpits for selected national populations.

ANTHROPOMETRIC DIFFERENCES AMONG VARIOUS POPULATIONS

Unfortunately, anthropometric data on all potential manufacturing and using populations are not readily available. Data on the populations of Central and South American and African nations are very difficult to find. Similarly, data on Near and Far Eastern populations are sparse, except for those of Iran (Noorani & Dillard 1971), India (Mookerjee & Bhattacharya 1956), Korea (Kay 1961), and Japan (Oshima, Fujimoto, Oguro, Tobimatsu, Mori, Watanabe, & Alexander 1965). Some Southeast Asian nations, particularly Thailand (White 1964*a*) and Vietnam (White 1964*b*), have been well studied. The Australian population has also been reasonably well described (Aird, Bond, & Carrington 1958; Carrington 1959). Good data are available for most North American (Alexander, Garrett, & Flannery 1969; Alexander & Laubach 1968; Ashe, Roberts, & Bodenman 1943; Clauser, Tucker, McConville, Churchill, Laubach, & Reardon 1972; Climatic Research Laboratory 1947; Daniels, Meyers, & Worrall 1953; Freedman, Huntington, Davis, Magee, Milstead, & Kirkpatrick 1946; Garrett 1968, 1970*a*, 1970*b*, 1971*a*, 1971*b*; Gifford, Provost, & Lazo 1965; Hertzberg, Daniels, & Churchill 1954; Hertzberg, Emanuel, & Alexander 1956; O'Brien & Shelton 1941; Randall & Baer 1951; Randall, Damon, Benton, & Patt 1946; Snow & Snyder 1965; White 1961) and West European populations (Ducros 1955; Evrard 1954; Hertzberg, Churchill, Dupertuis, White, & Damon 1963; Laboratoire d'Anthropologie 1965; Morant & Whittingham 1952; Udjus 1964), except for those of Mexico, Spain, and Portugal. Among East and Southeast European countries, Bulgaria (Bulgarian Academy of Sciences 1965) and Greece (Hertzberg et al. 1963) have been best studied. Anthropometric data on Russian populations are not complete. Anthropometric data from most available international sources have been collated by Garrett and Kennedy (1971).

A serious deficiency of almost all anthropometric data is that they have been obtained primarily for military populations. In general, the civilian populations of the world have been neglected.

What are the principal anthropometric differences among the populations of the world? For design purposes, are the differences significant? Are some differences irrelevant to designers? To answer these questions, it is instructive to consider some differences among various national military populations.

STATURE

Figure 1 shows comparative percentile values for stature for military populations of the United States (unpublished data), West Germany (unpublished data), France (Laboratoire d'Anthropologie 1965), Italy (Hertzberg et al. 1963), Japan (Oshima et al. 1965), Thailand (White 1964*a*), and Vietnam (White 1964*b*). The U.S. Air Force flight population is among the largest, anthropometrically, of all Western nations that manufacture and export aircraft. Since it is also one of the most thoroughly studied populations, I have used it as a standard against which to compare the others. The percentile values on the left in Figure 1 are for U.S. flight personnel and are based on unpublished data gathered in 1967. The other countries are arranged from left to right in descending order of their values for the 50th percentile in stature.

It is customary in the United States Air Force to design for the central 90 percent of the population. In the case of stature this means from the 5th percentile (167.3 cm) to the 95th percentile (187.7 cm), a design range of 20.4 cm. The 5 percent of the population above the 95th percentile and the 5 percent below the 5th percentile would not be specifically accommodated. This does not mean that 10 percent of the population is necessarily excluded. Fortunately, in a great many cases, the human body can make some accommodation to an inadequately designed system. There are, however, some instances in which very nearly all those not specifically accommodated will not be able to operate the aircraft safely. Examples are the reach distance to critical hand controls, ejection clearance, rudder pedal distance, and seat-to-canopy distance.

The United States, Germany, France, and Italy are representative of the North American and Western European populations where we find the

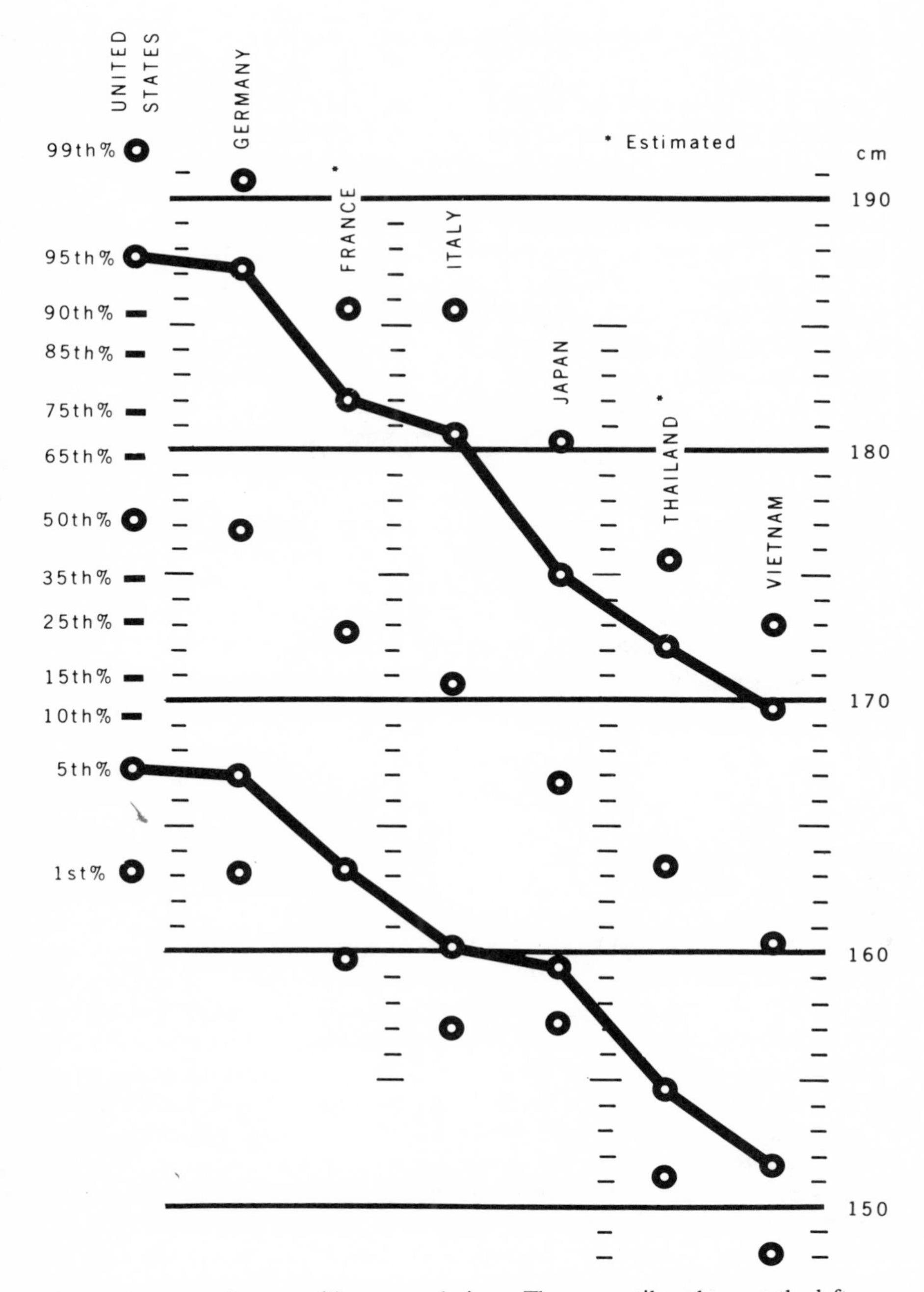

Fig. 1. Statures of seven military populations. The percentile values on the left are for U.S. Air Force flight personnel. The interval between the 1st and 5th percentiles should be approximately equal to the interval between the 95th and 99th percentiles. Since they are not in the case of the Japanese data, there is some question about the validity of these 1st and 99th percentiles.

tallest of the industrialized peoples. Thailand and Vietnam represent those nations where we find the smallest of the industrialized peoples. Comparisons among these groups are startling to the designer. For instance, the 5th to 95th percentile design range for Americans would accommodate essentially the same percentage of Germans (6th to 96th percentile), but only approximately the upper 80 percent of the French (19th to 99th percentile), 69 percent of Italians (30th to 99th percentile), 43 percent of Japanese (57th percentile to the top of the range), 24 percent of Thai (76th percentile to the top of the range), and 14 percent of Vietnamese (86th percentile to the top of the range). On the other hand, if we use the French 5th to 95th percentile range for design purposes, we would accommodate approximately 77 percent each of Americans and Germans (1st to 78th percentile) 83 percent of the Italians (13th to 96th percentile), 76 percent of the Japanese (24th percentile to the top of the range), 51 percent of the Thai (49th percentile to the top of the range), and about 30 percent of the Vietnamese (70th percentile to the top of the range).

Notice the position of the Japanese population. The 50th percentile for stature is almost exactly midway between that for Italy and Thailand. The 95th percentile Japanese is only slightly shorter than the 99th percentile Thai, and the 95th percentile Japanese is equivalent to the 76th percentile Italian.

SITTING HEIGHT

Figure 2 gives data on the sitting heights of the same populations as those in Figure 1. Sitting height is a far more critical dimension in laying out the geometry of the aircraft cockpit, since it must be considered in determining the depth of a cockpit, that is, the distance from the underside of the canopy to the heel-rest line (HRL). We see in Figure 2 a relationship among the various populations similar to that for stature in Figure 1. One prominent exception is that for sitting height the Japanese are slightly taller than the Italians and very nearly as tall as the French. For stature (Fig. 1), the Japanese are significantly shorter than both the Italians and the French. This raises a new consideration that must be taken into account when a Western (or non-Japanese) engineer is designing a cockpit to accommodate the Japanese user. Whereas almost all other populations are, more or less, scale models of each other, the Japanese are not. Japanese torsos are proportionately longer than their legs, as compared to most other populations. This implies, for instance, that an aircraft designed by the French for French use could meet the cockpit depth requirements for the Japanese, but would probably not be adequate in terms of rudder pedal distances. In general, a French aircraft could be adapted to the Japanese with fewer changes than could one designed by and for Americans or Germans.

THE RATIO OF SITTING HEIGHT TO STATURE

Differences among the proportions of various populations are more clearly illustrated in Table 1, which gives ratios between mean sitting height and mean stature among selected populations. This ratio is a measure of body proportions. Populations at the top of the list have proportionately longer legs, those at the bottom, proportionately shorter legs. The ratios for the populations selected fall naturally into three groups. To some extent the gaps in Table 1 are artifacts of selection that may not exist, or that may at least be greatly diminished, in actual fact. However, the separation between the Turks and the Japanese is unquestionably significant. Japanese body proportions are significantly different from those of most other populations, and this makes interchangeability of equipment with the Japanese very difficult.

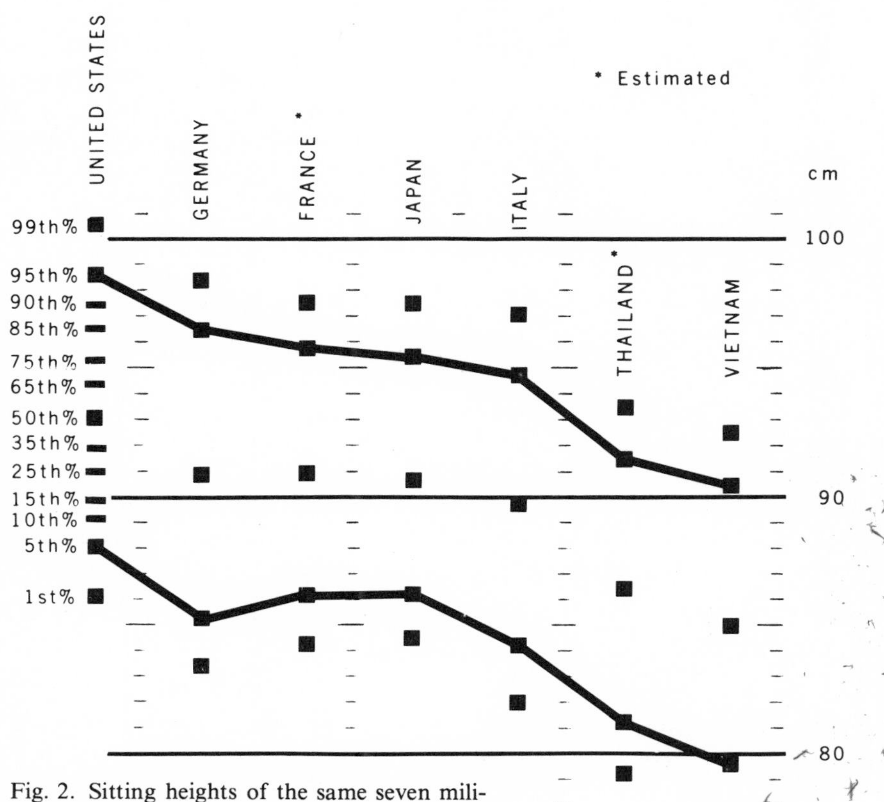

Fig. 2. Sitting heights of the same seven military populations as in Figure 1. The percentile values on the left are for U.S. Air Force flight personnel.

Table 1. Ratios of mean sitting height to mean stature for various national groups

	Men	Women		Men	Women
Germany (military)	0.514		France (military)	0.526	
Bulgaria (civilian)	0.522[a]	0.528[a]	United Kingdom (military)	0.526	
United States (civilian)	0.522	0.529[a]	Korea (military)	0.528[a]	
Norway (civilian)	0.522		Thailand (military)	0.529[a]	
Canada (military)	0.525		Greece (military)	0.529	
United States (military)	0.525	0.528[a]	Vietnam (military)	0.530[a]	
France (civilian)	0.526		Turkey (military)	0.530[a]	
Italy (military)	0.526		Japan (military)	0.544[a]	

[a]Indicates those populations for which the mean stature is less than the 20th percentile of the U.S. military population.

Those populations whose mean stature is smaller than the 20th percentile USAF stature are indicated by a superscript. These discrepancies are important in designing personal equipment, for example, pressure suits, a problem that will be discussed later in this paper. To anticipate, however, fitting problems are successively greater among those populations toward the bottom of the list and are magnified with a population such as the Japanese, who are not only differently proportioned but also quite different in overall body size.

APPLICATIONS TO COCKPIT DESIGN

To determine the effect of these anthropometric differences on cockpit design, let us assume that the USAF military standards for the basic dimensions of the cockpit are optimum for the population of U.S. pilots. While this assumption may be argued, the standards provide us with accepted cockpit geometries from which we can determine the changes necessary to accommodate other populations. Figure 3 gives the essential dimensions taken from one of these standards, Military Standard 33574, concerned with stick-controlled, fixed wing aircraft cockpits (Department of the Air Force 1969).

BASIC AMERICAN COCKPIT GEOMETRY

The basic reference points used generally throughout the American aircraft industry are the design eye position (DEP) and the neutral seat reference point (NSRP). The DEP may be taken as the primary reference point, since for a pilot to be properly positioned within the cockpit, he must raise or lower his seat so that his eyes are brought to the horizontal-vision line. The DEP is the average for all such positions along this line. The aircraft is designed so that when a pilot's eyes are brought to the DEP, he will have the best overall vision out of his aircraft, toward his instrument panel, and to special sighting devices and visual displays. At this adjustment he also has the minimum acceptable helmet clearance beneath the canopy. The DEP is located 25 cm perpendicularly from the back tangent line (BTL) and 80 cm above the NSRP.

The NSRP is a point within the cockpit and is part of the cockpit, not of the ejection seat. It is that point to which the ejection seat is adjusted during installation. The NSRP coincides with the Seat Reference Point (SRP) of the seat when the latter is at its midpoint of vertical adjustability. The SRP of the seat is defined as a point in the midline of the seat at the intersection of the depressed seat cushion and seat back. For positioning the pilot in the seat, it is convenient to consider that the pilot also has a SRP, when seated. The pilot's SRP is defined in essentially the same way as is that for the seat.

American engineers generally consider it easier to use this system of seat reference points than the DEP. Their reasoning is as follows: The NSRP is a fixed point within the cockpit; it is not the average of a series of points. When the midpoint of vertical movement of the SRP of the seat and the SRP of the operator are brought into coincidence with the NSRP, the NSRP is tied to structure as well as to the operator. On the other hand, the

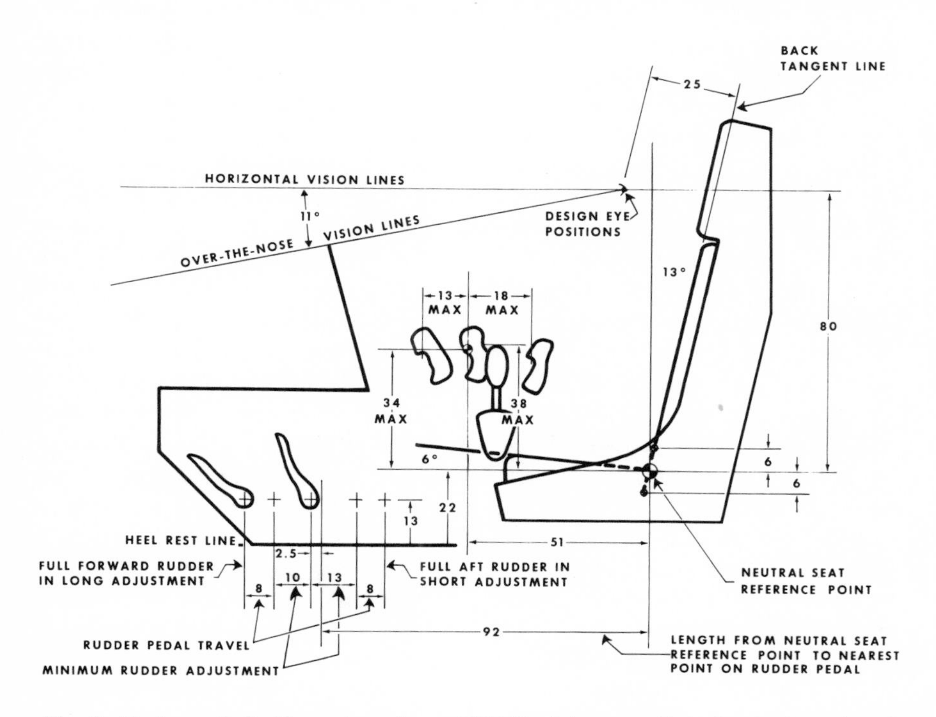

Fig. 3. Basic cockpit dimensions for a USAF stick-controlled, fixed-wing aircraft according to Military Standard 33574.

DEP is a point in space, unconnected to aircraft structure. In addition, as the pilot moves around in his seat during normal operation of the aircraft, his personal SRP remains reasonably stationary relative to the seat SRP and to the NSRP, that is, his own SRP moves within a relatively small envelope. His eyes, of course, move through a very much larger envelope. The SRP reference system is not, by any means, perfect. However, it is much more desirable from the designer's standpoint if hand- and foot-operated controls can be referenced to a point that remains relatively stationary.

When the pilot raises his seat, he raises his own and the seat's SRPs above the NSRP, carrying him away from the rudder pedals. Lowering the seat has the opposite effect; it brings him closer to these controls. The positions of the rudder pedals must, therefore, be adjustable fore and aft to compensate for such movements. In Military Standard 33574, the rudder pedal reference point is 92 cm forward of the NSRP and 13 cm above the heel-rest line. The latter is 22 cm below NSRP. Rudder pedal adjustability of approximately 10 cm forward and 13 cm aft is considered adequate. The standard further recommends that fore and aft rudder pedal movement during operation be limited to about 8 cm in each direction.

Since the control stick and other hand-operated controls do not move with the seat, they should be made adjustable or should be located so that they can be conveniently operated at all seat adjustments. In actual fact, hand-operated controls in American aircraft of this type are not made adjustable. The control stick reference point is generally located at a maximum of 34 cm above and 51 cm forward from the NSRP. This is sufficiently high in the cockpit and sufficiently close to the operator so that the stick can be conveniently reached by a high percentage of our population at any seat position. At the same time the stick is sufficiently far away to provide adequate clearance between the pilot's thighs and to provide full travel throughout its full movement envelope. The episode recounted at the beginning of this paper illustrated the consequences of not providing sufficient clearance between the pilot's thighs for full control movement. Hand controls other than the stick are located in reach zones dictated by their importance and the probable level of restraint of the pilot during operation.

In our high performance aircraft with ejection seats, we require that the seat be adjustable 6 cm above and below the NSRP along a line approximately parallel to the back tangent line, but exactly parallel to the ejection line. This requirement is necessary to maintain the seat-and-pilot center of mass at a constant distance from the ejection line. However, it introduces a very severe design problem in these aircraft. The very pilots who find it necessary to raise their seats to reach the horizontal-vision line are those of shorter torsos and, usually, correspondingly shorter limbs. Since the adjustment is invariably up and away from (or down and toward) his controls, it forces the smaller pilot *away* from and the larger pilot *toward* the aircraft controls, the exact opposite of what is required. Further discussion of this aspect of cockpit design would take us far afield from the immediate subject. It is, however, an extremely disconcerting feature of cockpit-design requirements. Multiposition cockpits of larger, slower aircraft, not equipped with ejection seats, generally provide both vertical and fore and aft seat adjustability, so the pilot can position himself more appropriately in the cockpit.

ADAPTATIONS OF THE AMERICAN COCKPIT TO OTHER NATIONALITIES

Figure 4 illustrates some of the basic dimensional changes that would be required in this American standard to accommodate the cockpit to the requirements of the Japanese and Vietnamese. The changes are drastic ones. The 13° back angle, 6° seat angle, and 22 cm heel-rest line are maintained without change. Because of the smaller sitting eye height among both the Japanese and Vietnamese, however, it would be necessary to lower the horizontal-vision line (and DEP) by approximately 3 cm for the Japanese and 7 cm for the Vietnamese, to 77 and 73 cm respectively, above the NSRP. It might also be possible to move this point one or two cm aft, since head length for these populations is somewhat shorter than for Americans. The latter design change would depend to some extent on their sitting posture and would require that special high-altitude helmets be designed to accommodate their generally smaller heads.

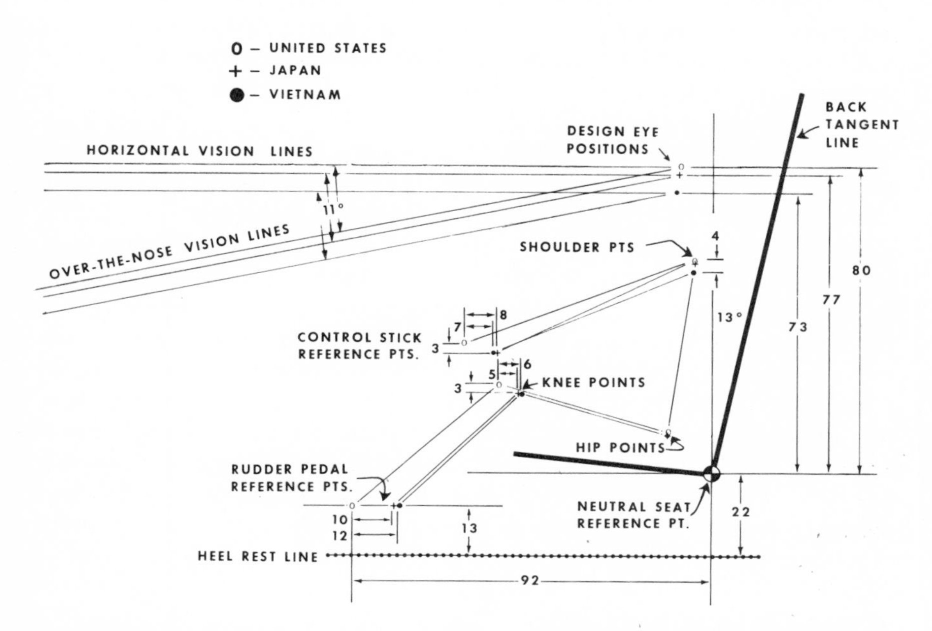

Fig. 4. Modifications in the basic cockpit geometry of Figure 3 required to accommodate Japanese and Vietnamese pilots.

The shoulder point is important for determining the placement of hand-operated controls. This point would undoubtedly be lower for Japanese than for American pilots, but the available anthropometric data are not sufficiently detailed to permit a precise estimate of how much lower. For the Vietnamese this point would be approximately 4 cm lower than for Americans. Again, it might be possible to put both points somewhat to the rear of that for American pilots, depending on the posture of both groups.

The altered shoulder points and shorter reach for these populations make it necessary to move the control stick aft about 8 cm for the Japanese and 7 cm for the Vietnamese. Since the Japanese have longer torsos and a higher shoulder point than the Vietnamese, but have roughly comparable arm lengths, the control stick for the Japanese cockpit would have to be moved slightly further to the rear. All hand-operated controls other than the control stick would require movement closer to the operator by similar amounts, that is by 8 cm and 7 cm, respectively, for the Japanese and Vietnamese.

The hip point for Americans is approximately 10 cm above and 12 cm forward of the SRP for the seat. The hip point for our other two populations is unknown. For purposes of this illustration, I have placed this point 1 cm lower than that for Americans.

The Japanese and Vietnamese are quite a bit shorter than Americans in the length of their legs, and the distance to which their knees extend forward is correspondingly smaller by approximately 5 cm for the Japanese and 6 cm for the Vietnamese. This means that rudder pedals must be moved to the rear 10 and 12 cm respectively. Since in both cases knees do not rise as high as American knees, the control stick could also be moved downward approximately 3 cm.

At first glance, these changes in basic cockpit dimensions may appear negligible. Their overall effect, however, is to produce a significantly shorter and shallower cockpit. Military Standard 33574 specifies that the distance from the underside of the canopy to the heel-rest line should be at least 127 cm. The changes discussed so far permit the canopy distance to be reduced by at least 3 cm for the Japanese and by approximately 7 cm for the Vietnamese. In addition, the shorter legs of both groups would necessitate a reduction in the distance from the NSRP to heel-rest line of about 4 cm, from 22 to 18 cm. The net result of all these changes would be a reduction in required cockpit depth to 120 cm for the Japanese and to 116 cm for the Vietnamese. Any aeronautical engineer would be extremely happy if he were told he could reduce the depth of his cockpit to these amounts, because this one change could bring about a significant increase in aircraft performance.

REACH ENVELOPES

Reach capability to hand-operated controls is obviously an important aspect of cockpit layout. It is often very difficult, even under the best conditions, to find sufficient space in suitable locations for the placement of controls. The factors involved are several and usually conflicting. When the control has an associated visual display, the display and control should be close to one another. As the need for more displays and controls increases, suitable space for their location is at a premium. Miniaturization and integration have provided some relief, but the spatial problems are far from being solved. If reach and visual capabilities permitted, a greater number of instruments could be handled by moving the instrument panel farther away from the pilot. However, since displays and controls must be seen and read as well as reached, there are practical size and distance limitations that must be observed.

For the American population, we can describe in three dimensions the reach envelopes within which controls may be placed and within which controls can be reached by 95 to 99 percent of the population (Kennedy 1964), with and without the effects of encumbering personal equipment (Garrett, Alexander, & Matthews 1970). For the military population of India, such an envelope would have to be reduced in size by about 6 percent; for Koreans, 13 percent. Some other populations would require even greater reductions. Figure 5 illustrates the horizontal contours predicted for these envelopes at the 38-cm (15-inch) level above the SRP. The 5th percentile reach forward at this level is 67 cm for Americans, approximately 63 cm for Indians, and 59 cm for Koreans. The 5th percentile reach to the right at this level is 81 cm for Americans, approximately 76 cm for Indians, and 70 cm for Koreans. In aircraft of equivalent complexity and performance, the requirement to place controls and displays closer reduces the available space even further. The closer controls must be placed to the pilot, the less surface there is available on which to mount them.

To appreciate how this problem has increased through the years, some single-seat aircraft of World War I had only three controls that the pilot had to reach when he was firmly strapped in the cockpit: the control stick, fuel cocks, and machine gun trigger. Table 2 shows the controls that must now be placed within the primary reach zone of modern military aircraft.

SIZE AND SHAPE OF HAND CONTROLS

The size and shape of hand controls will also require change if they are to be used by foreign military populations. Figure 6 illustrates the USAF standard hand grip, the MC-2 stick-control grip, used almost universally for stick-controlled, fixed-wing aircraft. Some American pilots feel that this hand grip should be somewhat smaller, so that all the switches could be operated without changing the position of the hand on the grip. If American pilots have difficulties operating some of these switches, one can expect these difficulties to be magnified among using populations with smaller hands. The ability to operate these switches seems to be associated with hand length. The longer the hand, the more easily the switches can be reached. Based on differences in hand length, the Turkish,

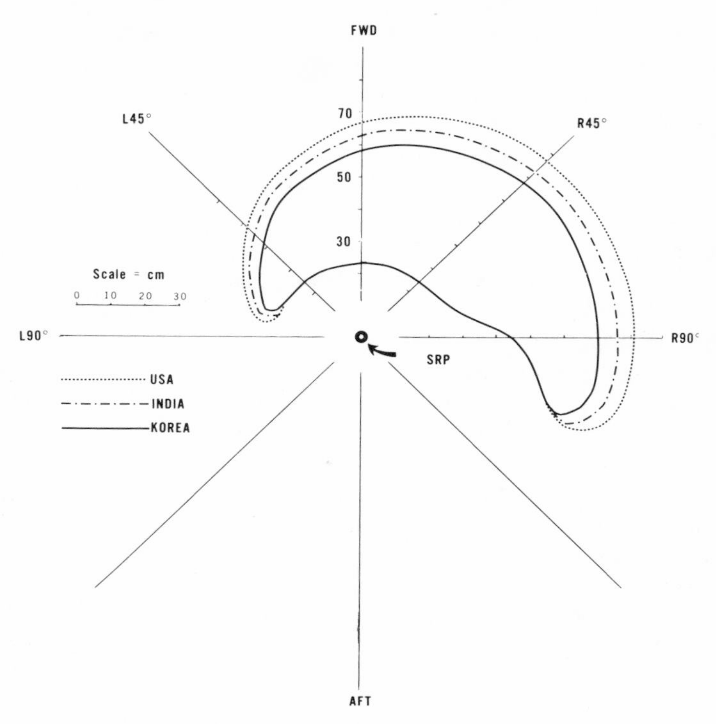

Fig. 5. Horizontal contours of the reach envelopes for three populations 38 cm above the seat reference point.

Greek, and Italian military might require about a 2 percent reduction in overall size of the hand grip; the Japanese and Korean, 4 percent; and the Vietnamese, 8 percent. At what point these reductions become significant is a matter for argument. Similar reductions would be required for controls on the power quadrant, especially since it is often used to mount critical switches.

Although there is a great amount of personal preference involved in the design and acceptance of hand-held multiple controllers, the control situations in which they are used are multiplying. Figure 7 illustrates some currently being manufactured by one company. An overriding consideration governing the size of such multiple controllers is that all switches are enclosed within the body of the grip. Thus, the limit below which a grip may not be reduced in size is determined by the number of switches and the degree to which they can be miniaturized.

PERSONAL-PROTECTIVE EQUIPMENT

The designer of personal-protective equipment, as well as the cockpit designer, must provide for the different body-size requirements of foreign

Table 2. Hand controls that must be placed within the primary reach zone in modern military aircraft[a]

Control stick/wheel	Life-support controls
Power quadrant	Ejection controls
Fuel mixture	Shoulder-harness lock
Propeller speed	Emergency and normal landing gears
Propeller feathering	Arresting-hook drop
Wing sweep	Auto-pilot disconnect
Speed brakes	Bail-out/ditching alarm control
Flaps	Microphone switch
Emergency trim	Emergency electrical power controls
Emergency brakes	Fire extinguishing and shut-off
Emergency canopy jettison	Emergency engine shut-down
Emergency external stores jettison	Primary fuel selector

[a] The pilot is assumed to be sitting with his shoulders back and with his shoulder harness locked.

users. Items such as full- and partial-pressure high-altitude garments, anti-g suits, flight coveralls, and parachute harnesses, play an extremely important role in the overall compatibility between the pilot and his crew station. At least three aspects of equipment design must be considered: (1) the accommodation or fit of the item of equipment to the operator, (2) the degree to which it provides him with the physiological protection intended, and (3) the compatibility of the personal–protective equipment with the layout of the cockpit.

If a foreign-using population is not greatly different anthropometrically from the population for which an article was sized, the problem of fit can be alleviated but not necessarily solved by altering the tariff for the individual sizes. For instance, the USAF height-weight sizing systems (Emanuel, Alexander, Churchill, & Truett 1959) based on data from Hertzberg et al. (1954) for full-body garments such as full- and partial-pressure suits, anti-g garments, and flight coveralls, can fit a large percentage of Italian military personnel between the 5th and 95th percentiles for height and weight. While the USAF population requires more of the sizes in the center of the range (the medium-sized suits), the Italians need more of the smaller sizes and few, if any, of the largest sizes. When a population has measurements that are even closer to American measurements, it could be accommodated in a similar manner. This procedure, however, is less effective when the anthropometric differences from the USAF population are greater. With some populations, such as those of Bulgaria, Turkey, Korea, Thailand, and Vietnam, additional smaller sizes must be added and the largest sizes disregarded.

In addition to overall differences in body size, differences in body proportions play an important role in the interchangeability of personal equipment. For instance, American-made partial-pressure suits will not fit Japanese pilots. Unfortunately, the arms and legs of the American garments are much too long. In the case of the Japanese, a completely new sizing system had to be introduced to accommodate to their different body sizes and proportions.

The use of personal equipment in the cockpit has significantly increased the complexity of the design process. Pressure suits, for example, create a series of problems that have been among the most difficult to solve. Many of those problems can be traced to two characteristics of these garments: (1) they add a great amount of bulk to the body, and (2) they decrease the mobility of the arms and legs. Inflating the suit magnifies both problems. Increasing the volume of the cockpit to accommodate for increased suit volume and relocating emergency controls closer to the pilot to compensate for his reduced reach capability are diametrically opposed requirements.

If a pressure suit is not properly fitted, mobility, and, therefore, arm and leg reach, are reduced disproportionately. Bulk, of course, is excessive if the garment is too large. It follows that any effort to retrofit or redesign an aircraft cockpit to accommodate a foreign population of significantly different measurements must be systematic. It must deal with all elements of the cockpit, not just those that interface between the cockpit and the pilot.

CONCLUSION

I have presented representative examples of the kinds of problems an American aircraft designer can expect to encounter when he designs a cockpit for selected foreign military populations. I have also discussed

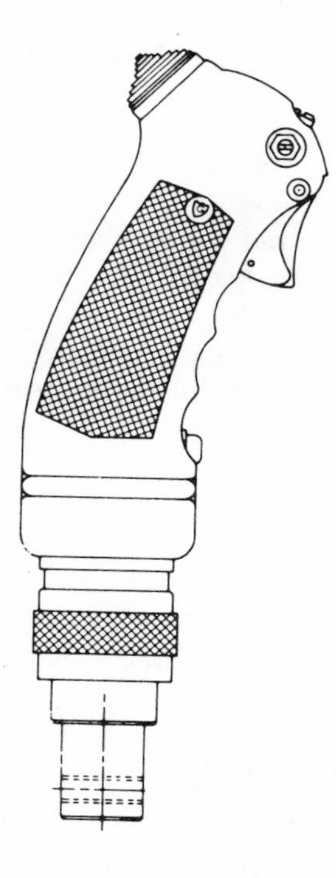

Fig. 6. The USAF MC-2 stick-control grip.

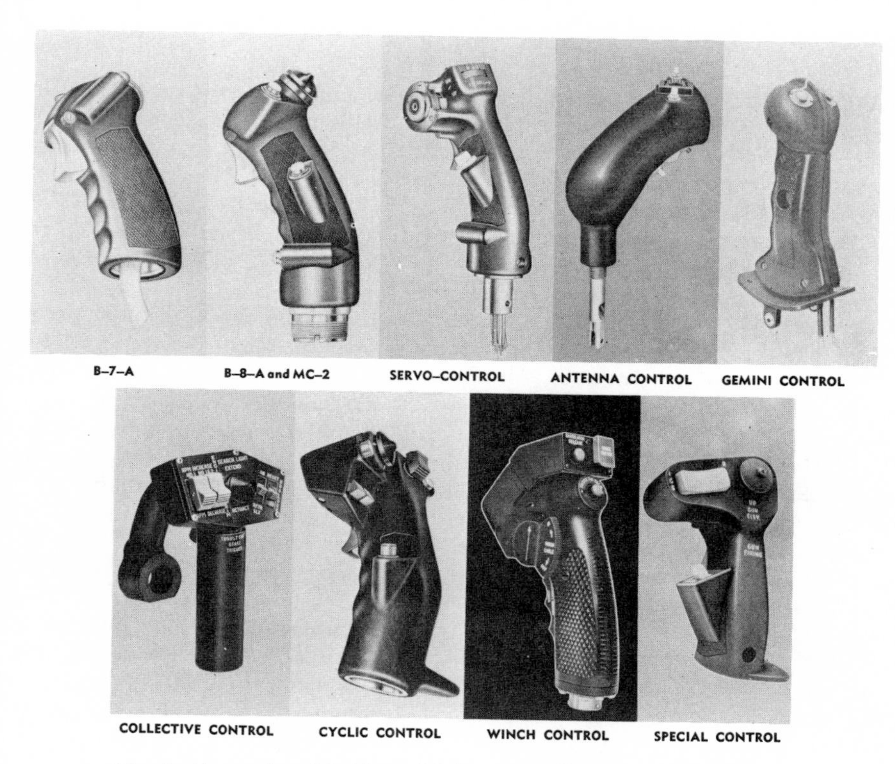

Fig. 7. Some hand-held multiple controllers currently being manufactured by one company.

ways in which some of these problems may be solved. The problems discussed here by no means exhaust those that the designer faces. Others are, unfortunately, difficult to anticipate because anthropometric data are incomplete for most national populations.

The spatial geometry of a cockpit designed for one population can and should be quite different from that designed for another. While this paper has concentrated on military aircraft, much of what has been said applies to civilian aircraft and to a great number of other work places that must fit the human operator, be they in vehicles, shops, or offices. Ideally, a nation should procure aircraft with cockpits specifically designed for its own population. Unfortunately, for reasons of economy, expedience, and performance, to say nothing of ignorance regarding the anthropometric characteristics of one's own national population, aircraft are shipped unaltered across many international borders. This practice usually leads to degradation in the performance of the aircraft as well as that of the pilot. The costs of such degradation in terms of human lives and money fully justify intensive efforts to correct the situation as it exists today.

REFERENCES

Aird, W. M., Bond, F. R., and Carrington, G. M. Anthropometric survey of male members of the Australian Army. Part I: Clothing survey. Report Number 3/58. Canberra, Australia: Army Operational Research Group, Australian Military Forces, June 1958.

Alexander, M., Garrett, J. W., and Flannery, M. P. Anthropometric dimensions of Air Force pressure-suited personnel for workspace and design criteria. Report Number AMRL-TR-69-6. Wright-Patterson Air Force Base, Ohio: Aerospace Medical Research Laboratory, August 1969.

Alexander, M., and Laubach, L. L. Anthropometric dimensions of the human ear (a photogrammetric study of USAF flight personnel). Report Number AMRL-TR-67-203. Wright-Patterson Air Force Base, Ohio: Aerospace Medical Research Laboratory, January 1968.

Ashe, W. F., Roberts, L. B., and Bodenman, P. Anthropometric measurements. Project 9, Report Number 741-3. Fort Knox, Kentucky: Armored Force Medical Research Laboratory, U.S. Army Medical Research Laboratory, February 1943.

Bulgarian Academy of Sciences, Institutes of Pedagogy and Morphology. *Physical development and fitness of the Bulgarian people from the birth up to the age of twenty-six. Volume I: Tables*. Sofia, Bulgaria: Bulgarian Academy of Sciences Press, 1965.

Carrington, G. M. Anthropometric survey of male members of the Australian Army. Part 2: Medical survey. Canberra, Australia: Australian Army Operational Research Group, Australian Military Forces, June 1959.

Clauser, C. E., Tucker, P. E., McConville, J. T., Churchill, E., Laubach, L. L., and Reardon, J. A. Anthropometry of Air Force women. Report Number AMRL-TR-70-5. Wright-Patterson Air Force Base, Ohio: Aerospace Medical Research Laboratory, April 1972.

Climatic Research Laboratory. Survey of body size of Army personnel, male and female. Report No. 1: Methodology and general considerations (women). Report Number 124. Lawrence, Massachusetts: Climatic Research Laboratory, Quartermaster Corps, July 1947.

Daniels, G. S., Meyers, H. C., and Worrall, S. H. Anthropometry of WAF basic trainees. Report Number WADC-TR-53-12. Wright-Patterson Air Force Base, Ohio: Aeromedical Research Laboratory, July 1953.

Department of the Air Force. Dimensions, basic, cockpit, stick-controlled, fixed wing aircraft. Military Standard MS-33574. Washington, D.C.: U.S. Government Printing Office, June 1969.

Ducros, E. Statistiques de biométrie médicale élémentaire relatives au personnel navigant de l'armée de l'air française. In *Anthropometry and human engineering*. AGARDograph Number 5. London, England: Butterworths Scientific Publications, 1955.

Emanuel, I., Alexander, M., Churchill, E., and Truett, B. A height-weight sizing system for flight clothing. Report Number WADC-TR-56-365. Wright-Patterson Air Force Base, Ohio: Wright Air Development Center, April 1959.

Evrard, E. *Statistiques biométriques relatives au personnel navigant de la force aérienne belge*. Belgium: Ministère de la Défense Nationale Force Aérienne, Direction du Service de Santé, May 1954.

Freedman, A., Huntington, E. C., Davis, G. C., Magee, R. B., Milstead, V. M., and Kirkpatrick, C. M. Foot dimensions of soldiers. Project Number T-13. Report Number S.G.O. 611. Fort Knox, Kentucky: Armored Medical Research Laboratory, March 1946.

Garrett, J. W. Clearance and performance values for the bare-handed and pressure-gloved operator. Report Number AMRL-TR-68-24. Wright-Patterson Air Force Base, Ohio: Aerospace Medical Research Laboratories, August 1968.

Garrett, J. W. Anthropometry of the Air Force female hand. Report Number AMRL-TR-69-26. Wright-Patterson Air Force Base, Ohio: Aerospace Medical Research Laboratory, March 1970*a*.

Garrett, J. W. Anthropometry of the hands of male Air Force flight personnel. Report Number AMRL-TR-69-42. Wright-Patterson Air Force Base, Ohio: Aerospace Medical Research Laboratory, March 1970*b*.

Garrett, J. W. An introduction to relaxed hand anthropometry. Report Number AMRL-TR-67-217. Wright-Patterson Air Force Base, Ohio: Aerospace Medical Research Laboratory, August 1971*a*.

Garrett, J. W. The adult human hand: some anthropometric and biomechanical considerations. *Human Factors*, 1971, **13**, 117–31. (*b*)

Garrett, J. W., Alexander, M., and Matthews, C. W. Placement of aircraft controls. Report Number AMRL-TR-70-33. Wright-Patterson Air Force Base, Ohio: Aerospace Medical Research Laboratory, September 1970.

Garrett, J. W., and Kennedy, K. W. A collation of anthropometry. Report Number AMRL-TR-68-1. Wright-Patterson Air Force Base, Ohio: Aerospace Medical Research Laboratory, March 1971.

Gifford, E. C., Provost, J. R., and Lazo, J. Anthropometry of naval aviators—1964. Report Number NAEC-ACEL-533. Philadelphia, Pennsylvania: Aerospace Crew Equipment Laboratory, U.S. Naval Air Engineering Center, October 1965.

Hertzberg, H. T. E., Churchill, E., Dupertuis, C. W., White, R. M., and Damon, A. *Anthropometric survey of Turkey, Greece, and Italy*. AGARDograph 73. New York: Macmillan, 1963.

Hertzberg, H. T. E., Daniels, G. S., and Churchill, E. Anthropometry of flying personnel—1950. Report Number WADC-TR-53-321. Wright-Patterson Air Force Base, Ohio: Aeromedical Research Laboratory, Wright Air Development Center, September 1954.

Hertzberg, H. T. E., Emanuel, I., and Alexander, M. The anthropometry of working positions: I. A preliminary study. Report Number WADC-TR-54-520. Wright-Patterson Air Force Base, Ohio: Aeromedical Research Laboratory, Wright Air Development Center, August 1956.

Kay, W. C. Anthropometry of the ROKAF pilots. *Republic of Korea Air Force Journal of Aviation Medicine*, 1961, **9**, 61–113.

Kennedy, K. W. Reach capability of the USAF population. Report Number AMRL-TDR-64-59. Wright-Patterson Air Force Base, Ohio: Aerospace Medical Research Laboratories, September 1964.

Laboratoire d'Anthropologie. Etude anthropométrique du personnel navigant français de l-aéronautique civile et militaire. Report Number Doc. A. A. 08/65. Paris, France: Laboratoire d'Anthropologie, Université de Paris, September 1965.

Mookerjee, M. K., and Bhattacharya, M. N. Body measurement in relation to cockpit design. *Journal of Aero Medical Society*, 1956, **3** (1), 32–37.

Morant, G. M., and Whittingham, D. G. V. A survey of measurements of feet and footwear of Royal Air Force personnel. Report Number FPRC 761. Farnborough, Hants, England: Royal Aircraft Establishment, July 1952.

Noorani, S., and Dillard, C. N., Jr. Anthropometric survey of the Imperial Iranian Armed Forces. Volume II. Statistical Data. Imperial Iranian Ground Forces Combat Research Evaluation Center, March 1971.

O'Brien, R., and Shelton, W. C. *Women's measurements for garment and pattern construction*. Bureau of Home Economics, Textiles and Clothing Division, Miscellaneous Publication Number 454. Washington, D.C.: U.S. Department of Agriculture and Work Projects Administration, December 1941.

Oshima, M., Fujimoto, T., Oguro, T., Tobimatsu, N., Mori, T., Watanabe, T., and Alexander, M. Anthropometry of Japanese pilots. Report Number AMRL-TR-65-74. Wright-Patterson Air Force Base, March 1965.

Randall, F. E., and Baer, M. J. Survey of body size of Army personnel: Male and female methodology. Office of the Quartermaster General, Research and Development Division, Report Number 122. Lawrence, Massachusetts: Quartermaster Climatic Research Laboratory, October 1951.

Randall, F. E., Damon, A., Benton, R. S., and Patt, D. I. Human body size in military aircraft and personal equipment. Army Air Forces Technical Report 5501. Wright Field, Dayton, Ohio: Air Materiel Command, June 10, 1946.

Snow, C. C., and Snyder, R. G. Anthropometry of air traffic control trainees. Federal Aviation Agency Report Number AM 65-26. Oklahoma City, Oklahoma: Office of Aviation Medicine, September 1965.

Udjus, L. G. *Anthropometrical changes in Norwegian men in the twentieth century*. Norwegian Monographs on Medical Science. Oslo, Norway: Universitetsforlaget, 1964.

White, R. M. Anthropometry of Army aviators. Environmental Protection Research Division, Technical Report EP-150. Natick, Massachusetts, June 1961.

White, R. M. Anthropometric survey of the Royal Thai Armed Forces. Natick, Massachusetts: U.S. Army Natick Laboratories, June 1964*a*.

White, R. M. Anthropometric survey of the Armed Forces of the Republic of Vietnam. Natick, Massachusetts: U.S. Army Natick Laboratories, October 1964*b*.

FIVE

Muscle Strength as a Criterion in Control Design for Diverse Populations

K. H. E. KROEMER

The ability to exert muscular strength, like other anthropometric descriptors, characterizes an individual as well as a population. Inseparably associated with muscle strength are such factors as body dimensions, handedness, and motion stereotypes. All these characteristics can vary significantly with the sex, age, and profession of an individual, and with culture, degree of civilization, and area of origin (nationality) of a population.

Information on strength is important to the ergonomist who designs a new man-operated system or evaluates existing equipment. The muscular capabilities of a user population determine, or should determine, the types of controls[1] selected, their arrangement, their modes of operation, and their control dynamics. Simple machinery and, of course, hand tools generally pose rather high demands on the user's strength. However, even sophisticated manned systems are controlled by hand- or foot-operated controls.

K. H. E. Kroemer, Head, Ergonomics Laboratory, Federal Institute for Occupational Safety and Accident Research, Martener Strasse 435, D 46 Dortmund-Marten, Federal Republic of Germany.

The author is grateful to J. M. Christensen and C. E. Clauser (Dayton, Ohio, USA), H. W. Juergens (Kiel, Germany), K. W. Kennedy (Dayton, Ohio, USA), A. R. Lind (St. Louis, Missouri, USA), H. Monod (Paris, France), H. Nutzhorn (Bremen, Germany), R. N. Sen (Calcutta, India), and M. J. Warrick (Dayton, Ohio, USA), for helpful critiques and information during the preparation of Technical Report 72-46 of the Aerospace Medical Research Laboratory (Kroemer 1974). That report, much shortened and revised, served as the basis for this article.

[1]In this paper the term "control" connotes any device serving to transmit muscular strength to the equipment.

Although some of these controls require very little energy input, many tax the operator's strength. In addition, critical controls, especially those for emergency operation, often require the application of large forces or torques. For normal operation, on the other hand, the human factors engineer selects an "optimal" value within the continuum of strength of the prospective user population.

This paper is concerned with the practical application of information on muscular strength, laterality, and movement stereotypes of various populations, and, in particular, with the operational importance of the magnitude of strength critical for control layout, the various aspects of strength required in control operation, and the biomechanics of strength. Finally, a procedure is described that the designer can use to adapt a product to population-strength characteristics. Several case reports illustrate some of the problems encountered in designing for strength.

STRENGTH IS SPECIFIC

Strength is highly specific (Laubach, Kroemer, & Thordsen 1972; Whitley & Allen 1971) and this specificity has several aspects:

- Different definitions, terminologies, procedures, and interpretations have in the past led to different "strengths." As a result, attempts have been made recently to eliminate the ambiguities and to standardize the assessment of strength (Kroemer 1970*a*, 1974).
- Strength describes the ability to exert force or torque to a given instrument (or control) at a given location, in a given direction. Strength scores obtained on different controls or under different conditions are often not highly intercorrelated.
- Strength depends not only on muscular capability, but also on health and training, technique and experience, motivation and body position.

Such considerations mean that an ergonomist must decide whether the available information on strength is pertinent to his design project. Specifically, he must decide whether the test conditions under which strength was measured resemble the design strength requirements sufficiently, and whether the test sample is representative of the prospective user population.

APPROACHES TO THE DESIGN OF CONTROLS

Every tool, device, equipment, or man-controlled system has its own purpose and must be designed to fulfill specific requirements. With respect to operator strength requirements, tools have to be grasped and manipulated; equipment has to be pushed, pulled, or lifted; and controls must be operated in prescribed ways, for example, rotated, pressed, and moved within given quadrants of the work space.

In the past, designers often started with the overall task of the system, equipment, or tool, and considered the human operator only after the "working side" of the machinery had been established. So, for example, machine tools, the power plant and the chassis of an automobile, or the performance requirements of an airplane were customarily determined first. Only then were attempts made to take characteristics of the human operator into account. By this time, the constraints imposed by the design left the designer very little freedom in the type of control he could select, its placement, or its mode of operation.

By contrast, human-oriented design starts by considering specific capacities of the operator, and improves, boosts, or modifies those characteristics through special devices. This is the way most hand tools were developed "naturally." It is fascinating to speculate what could be achieved if engineers were to start systematically with the operator, enhance his characteristics, and only then evaluate what the strengthened operator is able to do.

In any case, the capabilities of the operator and the requirements of the task must be matched from the very beginning of the design process. Such adaptation has become especially necessary because of the increasing complexity of contemporary man-operated systems and the growing reluctance of workers to tolerate inadequate working conditions.

MAGNITUDES OF STRENGTH CRITICAL FOR EQUIPMENT LAYOUT

In workplace layout, force and torque requirements must be adapted to the operator's strength available *under operational conditions*. Operator strength is critical in setting either a minimum value for control manipulation, so that even weak operators can actuate the control, or in setting a maximum value, so that the system will not be inadvertently actuated or damaged. Between the minimum and the maximum requirements is a "grey area" excluding (or including) certain portions of the user population. A given operator's strength depends on his muscular capabilities, which are subject to change depending on such things as his health, training, and state of fatigue, as well as on his skill in exerting his inherent capability, on the body support (reaction force) available to him, and on many other biomechanical and psychological variables. This article, however, will focus on differences in strength among diverse populations.

STRENGTH DATA FROM DIVERSE NATIONAL GROUPS

Cathcart, Hughes, and Chalmers (1935) reported an average force of 1,630 Newtons (N) exerted by approximately 10,000 British workers in isometrically pulling upward with both hands at a horizontal bar located at mid-thigh height. Using the same backlift action at a similarly located horizontal bar, 900 USAF cadets exerted a mean upward force of 2,340 N (Clarke 1945). A thorough examination of the original reports revealed no differences in experimental design, in experimental equipment, in measuring and recording techniques, or in statistical data treatment that might explain the discrepancies in these strength scores. Hence one has to accept as a fact that the cadets were indeed about one-half stronger than the workers. Similar background investigations are usually necessary when one is trying to match strength scores, either from different populations or from surveys conducted by researchers employing possibly different techniques.

Forces exerted on a rigid pedal with the preferred foot by attempted (isometric) extension of the leg, with the operator sitting and supported by a backrest, were measured by several researchers under a variety of experimental conditions (see Kroemer 1971*a*, for a compilation). Table 1 lists the largest average forces reported in each study. The means for each national population were computed from the sample means weighted for the number of subjects in the sample. Table 1 shows that the U.S. and British males yielded almost the same average values, but the German male subjects exerted only about half as much leg thrust. If the experimental procedures were indeed comparable in these studies, the data would indicate a distinct difference in exertable leg strength that should be taken into account by the designer. However, there are no anthropometric or biomechanical concomitants to make such a discrepancy plausible. Hence, the difference in strength scores in this case is likely to be an artifact of different experimental procedures.

Guthrie, Brislin, and Sinaiko (1970) report grip strength measurements taken on an Asian and a U.S. male population. Eighty-two Vietnamese young men were found to be considerably weaker than eighteen men from the United States. The 75th percentile grip strength of the former was about equal to the 25th percentile score of the latter. In this case, comparable samples were subjected to the same test by the same researchers to assess respective strength capabilities. This is the kind of data designers need to adapt equipment characteristics to various user populations. Unfortunately, such directly comparable data, taken systematically, are still hard to find in the literature.

The mean "maximal" weight of boxes, 66×28×18 cm in size, that have to be lifted with both hands 91 cm from the floor was found to be 54 kg by Emanuel, Chaffee, and Wing (1956). The mean "reasonable" weight of boxes, 30×30×15 cm in size, that have to be lifted with two hands 109 cm from the floor is reported to be between 32 and 44 kg by Switzer (1962). Finally, based on their own experimental findings as well as on a literature survey, Snook, Irvine, and Bass (1970) concluded that the maximum "permissible" weight to be lifted by unselected adult U.S. male workers is 23 kg.

Table 1. Mean maximal leg strength (in Newtons) for U.S., British, and German samples

	Source	Subjects	Average force	Standard deviation
U.S. males	Gough & Beard 1936	2 U.S. pilots	1,860	Not given
	Elbel 1949	515 U.S. pilots	2,520	440
	Martin & Johnson 1952	166 U.S. tank personnel	3,230	Not given
	Weighted mean	683 U.S. males	2,690	
British males	Hugh-Jones 1947	6 Englishmen "powerfully built"	3,770	Not given
	Hugh-Jones 1947	32 British soldiers	3,080	Not given
	Hugh-Jones 1947	16 British teenagers	3,070	Not given
	Rees & Graham 1952	20 British students	1,710	Not given
	Crawford 1954	5 British pilots	2,535	Not given
	Weighted mean	79 British males	2,750	
German males	Hertel 1930	11 German pilots and engineers	2,160	180
	Rohmert 1966	60 German men	1,402	300
	Weighted mean	71 German males	1,520	
German women	Müller 1936	2 German women	1,510	Not given
	Rohmert & Jenik 1971	10 German women	1,010	Not given
	Weighted mean	12 German females	1,100	

This example shows that values pertaining to maximal, tolerable, acceptable, reasonable, comfortable, and desirable efforts vary widely, depending on the work requirements and on the criteria used to find the "optimum" for each given case. They also vary considerably among different populations. Using the assumption that the load to be carried by a soldier should not exceed one-third of his body weight (White 1964*a*, *b*; Hart, Rowland, & Malina 1967), the equipment mass carried by Korean personnel should be limited to less than 20 kg, to about 18 kg for Thais, 17 kg for Vietnamese, and 24 kg for Americans (Anthropology Branch 1969).

IMPLICATIONS

These examples illustrate first that it is very difficult to tie together recommendations for the dynamic lifting of loads and to compare the lift loads with data on static strength capabilities as reported in the literature. Static strength scores are numerically much larger than the weights (mass forces) recommended for dynamic lifting or carrying, as discussed in detail elsewhere (Kroemer 1970*a*).

It is often assumed that a condition allowing maximum exertion of isometric forces is also optimal for other than static efforts. But "maximum" is the greatest quantity possible, the upper limit of variation, while "optimum" is the best, the most favorable condition. Hence, these terms refer to two different phenomena. Moreover, "optimum" needs further definition: optimum in what respect? With regard to muscular efforts, there are different optimal conditions, for example, for static forces in contrast to dynamic work, or for accurate movements in contrast to short outbursts of energy. In human engineering, an "optimal" work condition often is one in which the operator undergoes as little physical strain and fatigue as possible, so that he can perform his task for a long time without deterioration. What is "just not" straining or fatiguing depends, in a statistical sense, on the characteristics and capacities of the respective operator population.

Finally, these data illustrate the complexity of the problems encountered when one tries to compare strength data from different sources, referring to different populations, to be used for different design purposes. Published data cannot be accepted and applied uncritically.

MOVEMENT STEREOTYPES, HANDEDNESS, AND DEXTERITY

MOVEMENT STEREOTYPES

Certain populations exhibit strong and consistent expectancies about the kinds of equipment response that will follow from certain control actions. For example, automobile drivers expect a vehicle to move to the right if they turn the steering wheel clockwise. Such naturally expected relation-

Table 2. Stereotypes for control action—equipment responses in the US, Great Britain, and Germany[a]

Control action	Expected equipment response: Up		To the right		Forward		Clockwise		Start/on/more	
Up	√	US	Not	US			Not	US	√	US
	√	GB	Not	GB	?		Not	GB	Not	GB
	√	G	Not	G			Not	G	√	G
To the right	Not	US	√	US	Not	US			√	US
	Not	GB	√	GB	Not	GB	?		√	GB
	Not	G	√	G	Not	G			√	G
Forward			Not	US	√	US	Not	US	√	US
	?		Not	GB	√	GB	Not	GB	√	GB
			Not	G	√	G	Not	G	√	G
Clockwise	Not	US					√	US	√	US
	Not	GB	?		?		√	GB	√	GB
	√	G					√	G	√	G

√ =Established stereotype, recommended
Not =Not established, not recommended
? =Questionable or conditional association

[a]Adapted from Morgan et al. 1963; Kroemer 1967.

ships between control action and system response, existing without special training or instructions, are called population stereotypes.

Table 2 lists stereotypes relating four control actions with five equipment responses as established in the United States, Great Britain, and Germany (adapted from Morgan, Cook, Chapanis, & Lund 1963; and from Kroemer 1967, using DIN and ISO industrial standards). It is remarkable that almost all stereotypes listed are found in each of the three countries. One exception is the "Control Up-Equipment Start" relationship which does not apply in Great Britain but is accepted in the United States and in Germany. The other exception is the "Control Clockwise-Equipment Up" relationship, which is not established in both Anglo-Saxon countries, but is accepted and recommended in Germany, according to DIN 1410 for control operation and arrangement.

These data show that definite differences in stereotypical patterns exist even between populations closely related historically and culturally. Such differences should be taken into account in the design of equipment to be used in each of the countries. Although much more diversity might be expected in the stereotypes of less closely related countries, reliable information does not abound.[2] One may only speculate on how many false operations of equipment imported from another country occur daily, and how

[2]*Editor's note*: See the paper by Verhaegen et al. in this book.

many accidents may be attributable to mismatches of operator stereotypes and operational requirements.

HANDEDNESS AND DEXTERITY

Handedness, lateral preference, dexterity, laterality, and dominance are terms connoting types and degrees of a subject's ability to perform common tasks better with one hand or foot than with the other. However, dexterity also depends on the task to be performed. Different body segments may be used for, say, a finely controlled sensorimotor task, or for the exertion of brute strength (for a detailed discussion see, for example, Barnsley & Rabinovitch 1970; Palmer 1967). Using mainly questionnaires, Annett (1970) distinguished several patterns of preferences. These patterns were neither discrete nor restricted to certain categories of tasks, but were overlapping in several categories. Kimura and Vanderwolf (1970, p. 769) summarize the state of our knowledge as follows: "surprisingly little is known either about the nature of the motor skills involved in these (customarily executed) acts, or about the nature of the motor dominance described as hand preference." If the nature, definition, and assessment of laterality are still being debated, this may explain why we have such scant information on differences among various populations. At present, a rule of thumb is that less than 10 percent of any large national population is left-handed.

With respect to muscular strength, the evaluation of dexterity is relatively simple. The task of exerting a force to a measuring device is rather easily understood by almost any subject, and he can usually determine, by trial if necessary, with which hand or foot he prefers to exert the force. If laterality is not determined ad hoc, it can be assessed by comparing scores achieved with either hand separately. This approach uses measured performance rather than stated preference as the basis for determining laterality. Measures of grip strength are exceptionally simple, since one just has to squeeze a grip (Hunsicker & Donnelly 1955; Pangle & Garrett 1966; Schmidt & Toews 1970). Hence assessment of grip strength could serve to determine one type of strength handedness in international surveys.

The operator's ability to exert pressure, force, or torque under all operational or emergency conditions, in combination with his stereotypical and laterality patterns, is a critical factor for the designer. Muscular capabilities are structurally the same for all populations, although they may vary in magnitude according to training and health. However, stereotypical response and dexterity patterns may be fundamentally different among populations and may be rather difficult to change. Also, complicated interactions must be expected between stereotypes, learned responses, handedness, muscular training, and customary control operations about which we know mostly from personal experience or hearsay. Case 1 is an example of such information.

Case 1: A visitor from the USA rented an automobile in Japan. While driving in heavy traffic, requiring frequent gear shifting with the left hand, he found his left arm getting sore from the unusual exercise. Fatigued and unconsciously trying to avoid further strain, the visitor did not shift down in slowing traffic and stalled the engine when trying to accelerate rapidly in order not to be hit by another car. A serious accident resulted.

At present, no other choice is left to the designer of tools and equipment than to follow patterns established within certain industries or populations. In most countries, automobiles have the gear shift on the right, although in some countries the reverse is true. Single-seater airplanes usually have the throttle on the left. Sewing machines are commonly built "the wrong way," that is, designed so that the left-handed person can operate them best. "Leaving it the way it was" perpetuates accepted designs and styles and causes new products to conform to established procedures. It may also help to develop uniform strength patterns. Unfortunately, leaving it the way it was creates difficulties for international travel and commerce.

BIOMECHANICS OF STRENGTH IN RELATION TO CONTROL LOCATION

Deciding on an appropriate location for a control within the operator's reach envelope is one of the most critical features in design for human operation. The work space available for shirt-sleeve control operation has been described by Faulkner and Day (1970) and by Kennedy (1964). However, encumbering equipment or g-stresses can severely reduce the usable space. Overcrowding by too many controls can also become a serious problem (Kennedy 1972).[3]

Figures 1 through 5 show how human strength applications depend on the spatial relations between control and operator, and how the body support available to the subject affects the amount of force or torque he can develop. They also indicate a solution for the problem of how to design for populations with different strengths. Figure 1 demonstrates that maximal leg strength can be applied to a fixed pedal if it is located at about seat height and so far forward that the leg must be almost fully extended to reach it with the foot. Other arrangements of the pedal within the reach envelope of the leg reduce the force capacity.

Although differences in strength exist among individuals, there are no gross morphological differences in skeletal or muscular structure among different populations.[4] Hence, the relationships between force or torque and relative location of the measuring device should not change appreciably among populations. The generality of biomechanical principles must hold as long as the relations between such things as body dimensions, mechanical advantages, and pull angles are the same. For example, data show that while the absolute forces exerted by men and women differ

[3]*Editor's note*: See also the paper by Kennedy in this book.
[4]*Editor's note*: But see Table 6 in the paper by Roberts in this volume (page 26).

markedly (see Table 1), the relationships between location, body position, and exerted strength are similar for both sexes.

Figure 2 depicts body angles and body-support features necessary to describe the biomechanical relationships during exertion of leg and foot strength. Using these descriptors, Figures 3, 4, and 5 show the effects of changes in biomechanical variables on leg and foot strength. Similarly, Figure 6 schematically illustrates the torques exertable at different elbow angles. The magnitudes of elbow torques reported by about one dozen researchers from different countries vary considerably, probably reflecting variations in experimental techniques and test conditions. However, the relationship between strength and elbow angle is uniform.

The biomechanical interactions between (1) operator strength available at the control and (2) control location, are illustrated by the following application. Locating a pedal 100 cm directly in front of the seat reference point (SRP) allows approximately 95 percent of a male Central European population to operate this pedal with ease. However, this distance is distinctly too large for shorter-legged populations, such as many Asian males or European females. For male Koreans, for example, this pedal location requires approximately 50 percent of all operators to slide forward on the seat. In so doing they lose the firm support of the backrest, which in turn reduces their capability for strong leg thrusts on the pedal (anthro-

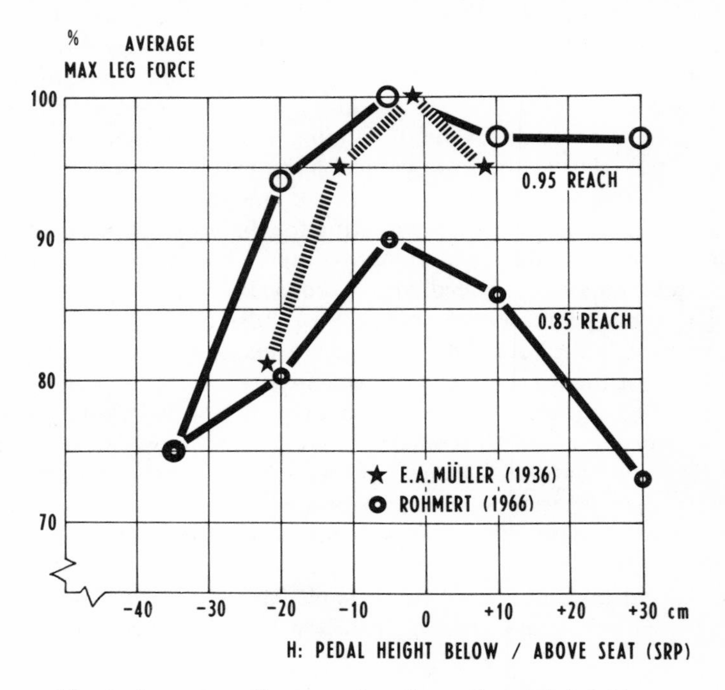

Fig. 1. Leg strength as a function of pedal height (Kroemer 1971*b*, 1974).

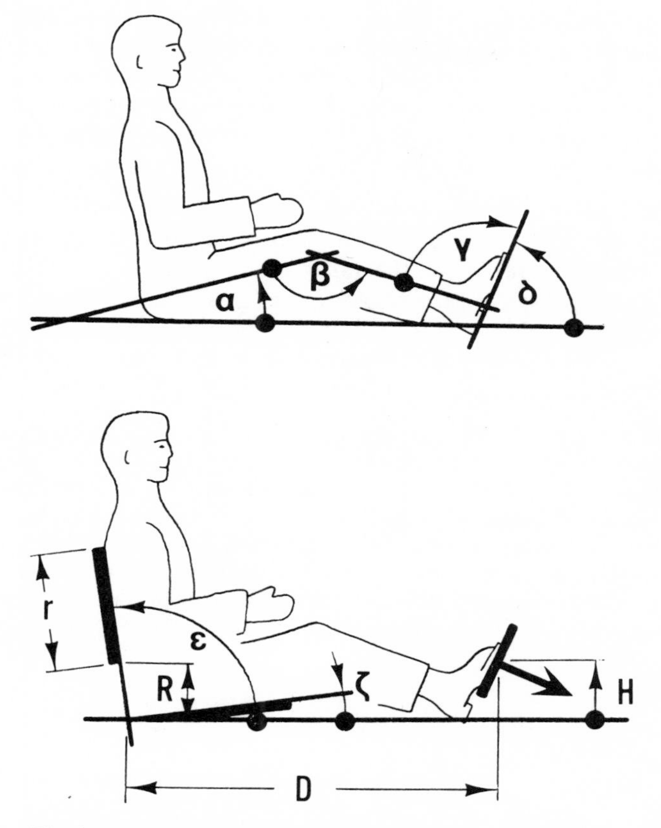

Fig. 2. Variables describing body posture and body support (Kroemer 1971*a*, 1971*b*, 1974).

pometric data from Garrett & Kennedy 1971). Figure 7 illustrates this condition. While the taller person exerts about the largest possible force at a knee angle of approximately 160° (see Fig. 4), the shorter operator attempting to reach the pedal has to move the lower part of his trunk away from the seat. Consequently, he receives supporting reaction force from the backrest at thorax height only, which, according to Figure 3, reduces his leg strength exertable at the pedal by at least 25 percent.

Case 2: In an aircraft, an important emergency control was located 55 cm above, 28 cm to the right of, and 70 cm in front of the SRP. These dimensions were chosen by the designer about 20 years ago so that the "average pilot"[5] could reach the control with the elbow at an angle of approximately 120°, and activate it by applying an upward force of at least 300N. An emergency arose during a test flight with a very small pilot at the controls. Pulled back into the seat by the automatic safety

[5]See Daniels (1952) for a discussion of the fallacy of the "average man" concept.

harness, the pilot could hardly bring his hand to the emergency handle, which was at the periphery of his reach. With increased g_z-acceleration pulling his extended arm down, the pilot could not develop the muscular strength necessary to activate the emergency control. Only exceptionally lucky circumstances prevented a serious accident. Control location and force requirements have been changed since to avoid such incidents.

Figure 8 illustrates the case just described. While the tall pilot can reach the control comfortably, the small operator can hardly reach the handle. Furthermore, while the taller operator, with an elbow angle of about 120°, can exert a rather large force, *F*, at the control (about 90 percent of his maximal strength; see Fig. 6), the small operator has to exert *F* with the arm extended, which reduces his strength capability to 50 percent or less of his maximum.

Case 3: Heavy American industrial machine tools have been delivered to India and are operated there by Indian personnel. Essential controls are arranged on the machines to be within reach of American workers. Indian operators, however, being generally of smaller size, have to stand on make-shift platforms or boxes to be able to reach some of these controls. In this disadvantageous biomechanical position, Indian operators strain themselves much more than do their American counterparts when applying the forces and torques necessary to operate the controls (personal communication from Dr. R. N. Sen 1971).

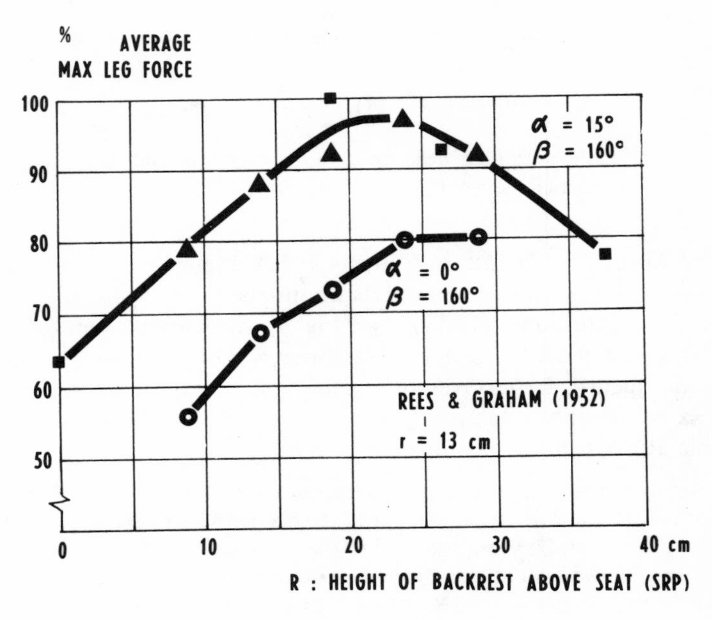

Fig. 3. Leg strength as a function of backrest height (Kroemer 1971*b*, 1974).

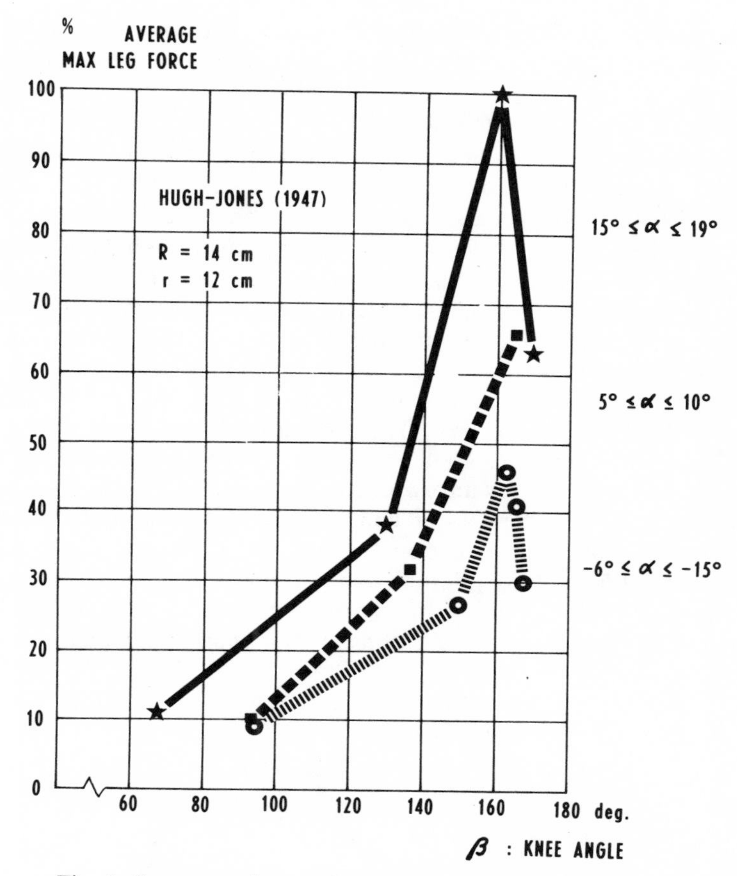

Fig. 4. Leg strength as a function of knee angle (Kroemer 1971*b*, 1974).

Figure 9 illustrates the situation when a small worker has to operate a control located at eye height for a 50th percentile U.S. male. Reaching up with the arms almost extended reduces force capability and endurance, increases energy expenditure, and so results in unnecessary strain and fatigue (Astrand & Rodahl 1970; Lehmann 1961; Scherrer 1967). Elevating the standing surface by a distance, *D*, about equivalent to the difference in eye heights between the average American and Indian, facilitates operation of the high control but simultaneously removes low controls from easy reach.

In each of the preceding examples, controls requiring muscular strength were located to suit a taller population, and consequently were outside the normal reach of operators from populations that are smaller in size. The reverse also happens, of course. Foot-operated controls in some automobiles designed for small-sized populations are too close to the seat and spaced too close together for drivers with longer legs and larger feet.

The solution to such design problems is to locate controls according to selected *ranges of joint angles and lengths of body segments* of user populations, instead of positioning controls at certain *linear distances from reference points*. When designing for ranges of body angles and segment lengths, operator strength available for control activation will vary relatively little among different populations, since such biomechanical relationships as joint angles, lever arms, mechanical advantages, and direction of g-stresses remain similar. Figures 1, and 3 through 6, illustrate how, by designing for body positions, the strength available at the control can be kept at the maximum.

To be sure, design according to body positions and proportions has been rather difficult in the past. The difficulties could be attributed in part to insufficient knowledge about biomechanical effects. In addition, it has been very difficult to take into account the complex interactions among biomechanical variables using conventional tables of anthropometric and strength data and conventional design tools. Finally, some designers apparently did not have relevant information or did not bother to use it adequately. Recent research in engineering anthropology and biomechanics (Reid 1973) and in the development of computerized biomechanical man-

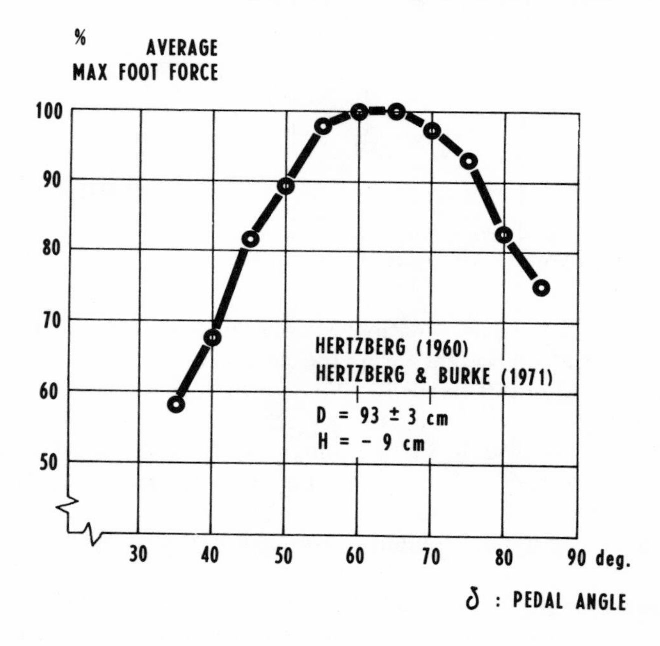

Fig. 5. Foot strength depending on pedal angle (Kroemer 1971*b*, 1974).

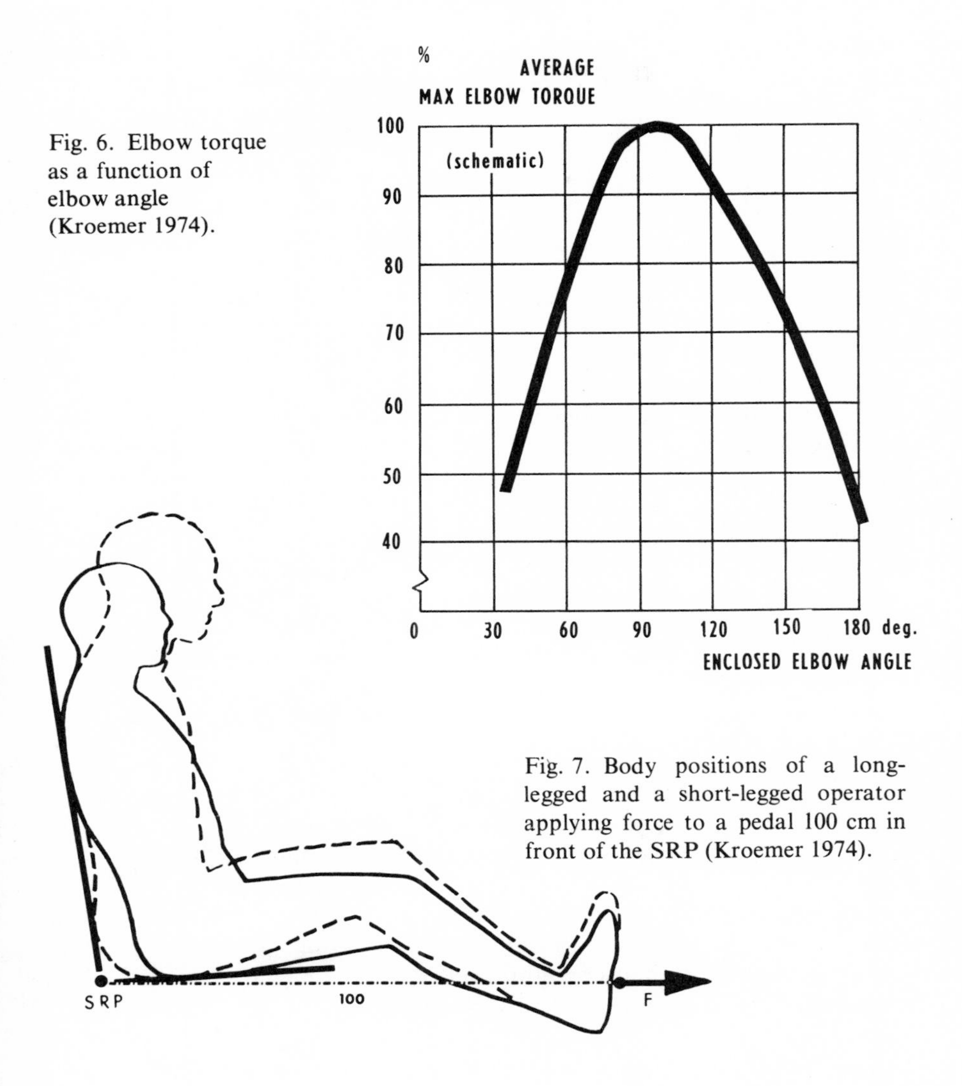

Fig. 6. Elbow torque as a function of elbow angle (Kroemer 1974).

Fig. 7. Body positions of a long-legged and a short-legged operator applying force to a pedal 100 cm in front of the SRP (Kroemer 1974).

models promises fast and easy access to such information and facilitates its use in the design process (Kroemer 1973). The Universities of Cincinnati, Florida and Michigan, the Boeing Company, and the Aerospace Medical Research Laboratory in the United States, the University of Nottingham in the United Kingdom, and other institutions, are actively developing computer models of man-workplace geometry that will enable the designer to use a computer terminal as an "advanced drawing board" in fitting new equipment to man's dimensions and biomechanical characteristics.

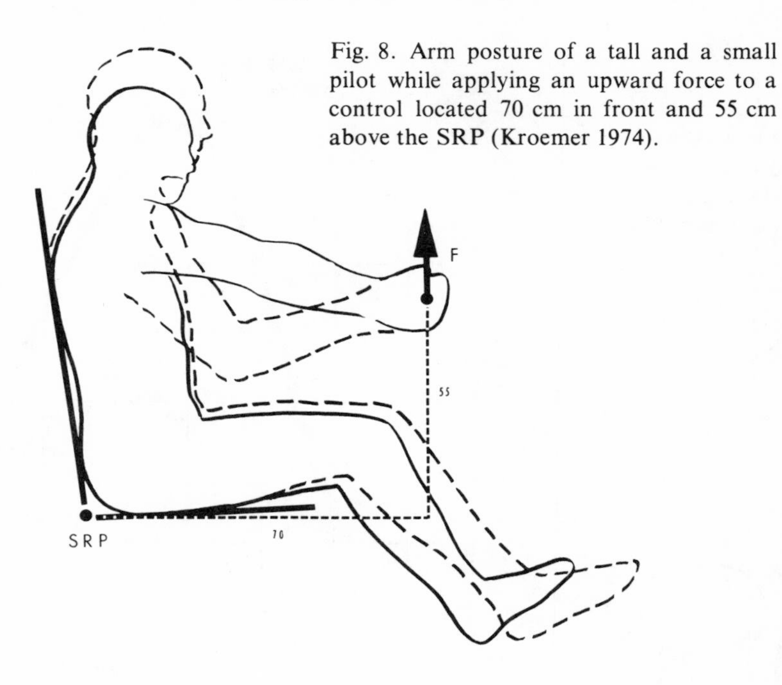

Fig. 8. Arm posture of a tall and a small pilot while applying an upward force to a control located 70 cm in front and 55 cm above the SRP (Kroemer 1974).

A PROCEDURE TO USE IN DESIGNING FOR OPERATOR STRENGTH

Following are some design procedures and rules that will enable the human factors engineer to design equipment according to the strength characteristics of different operator populations.

Step 1: Establish the critical anthropometric dimensions of the prospective user populations.

Step 2: Establish operational (minimal and maximal) reach envelopes for each distinct operator sample.

Step 3: Establish whether equipment worn, or environmental or other conditions can affect operator mobility and strength.

Step 4: Select for possible control locations those spaces in which the areas of convenient reach of all user groups overlap. If no such areas exist, establish what adjustments in control location and/or seat or standing platform are necessary to make reach spaces overlap. If no such adjustments can be made with reasonable effort, separate designs are unavoidable to suit different operator populations.

Step 5: Simultaneously,

- establish operator strength capabilities in the selected reach areas, considering operational conditions and preferred types of strength;
- establish system requirements in terms of control force or torque vectors (location, direction, magnitude).

Step 6: Match operator capabilities and system requirements. In particular, select a body posture and provide support so that the operator's body members are in comfortable, biomechanically advantageous positions. If the operator has to stand while working manually, his upper arms should be normally vertical or slightly elevated, and the elbow angle should be near 90 degrees. A sitting position is often preferable, in fact necessary, if controls must be worked with the feet frequently or con-

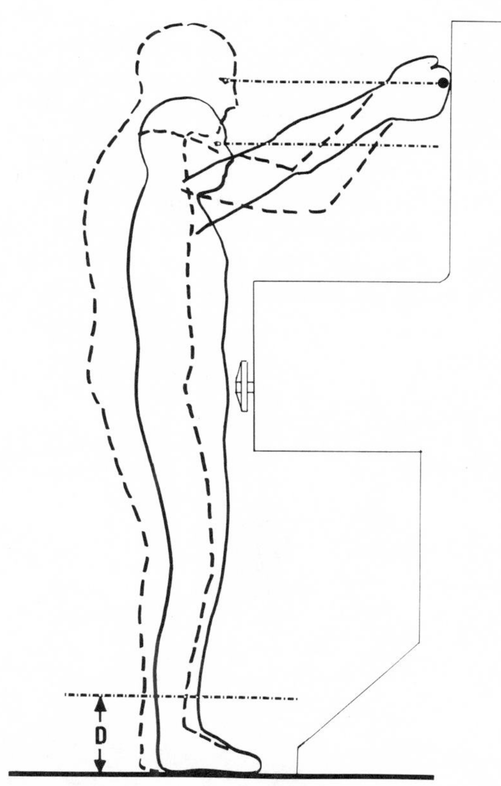

Fig. 9. Body posture of a small operator using heavy industrial machinery designed for taller operators (Kroemer 1974).

tinuously. For the exertion of rather large forward forces on pedals, the operator's legs should be nearly extended. For smaller force requirements, the knee angle can be between 150 and 100 degrees. The best arm position for the seated operator is the same as for the standing worker. (For further information, see Van Cott & Kincade 1972). Biomechanically advantageous ranges of body segment positions, in combination with anthropometric dimensions, determine the location, type, and operation of controls that require strength inputs from the operator.

Step 7: If a conflict arises in step 6, determine whether critical control force requirements are set by system requirements or by operator characteristics.

System requirements are usually either at a very low, or an extremely high level. A low level allows even very weak persons to operate the equipment. A high level will eliminate a large portion of prospective operators who cannot exert such strength. If the system requirement is exceedingly high, it must be lowered by redesigning the system. This can be done by using power boosters, such as power steering or fly-by-wire, or by changing the mechanical advantages by using different controls.

Operator strength values determine control characteristics according to what the designer considers to be optimal (tolerable, permissible, reasonable, or desirable) for the prospective operator. The optimum, the design strength value, is most conveniently expressed as a percentile value of operator strength. Since less than 20 percent of total strength can be maintained over practically indefinite periods of time without deterioration due

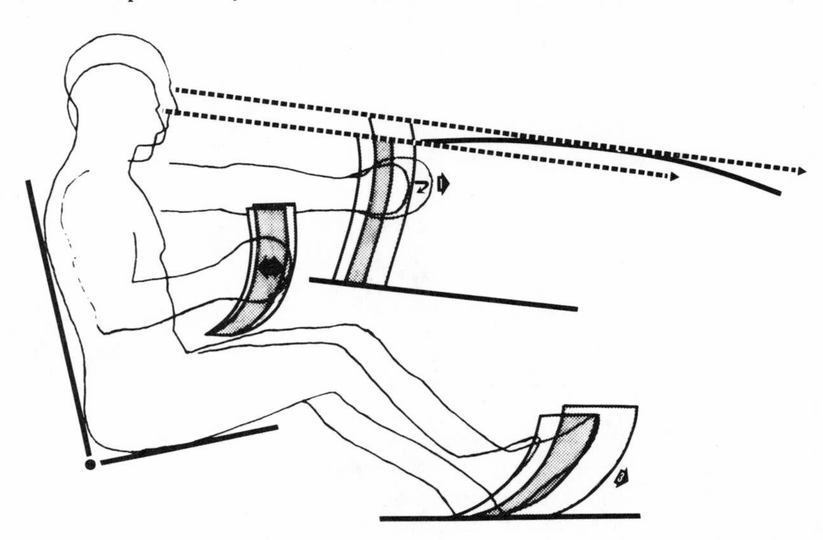

Fig. 10. Areas for the location of hand- and foot-operated controls for different user populations (Kroemer 1974).

to fatigue, the 20th percentile is an important cutoff value in design. Very often the human factors engineer has to set percentile ranges to exclude the extremely weak as well as the extraordinarily strong operators from his design considerations.

The procedures described in steps 1 through 7 apply primarily to new equipment that must be designed according to the biomechanical characteristics of different operator samples. However, the regimen also covers existing equipment that must be checked to determine its suitability for a new operator population. Figure 10 illustrates such a case.

An existing cockpit designed to fit a U.S. population was to be used by (a) Japanese operators, whose legs are shorter but whose trunk dimensions are similar, or by (b) Vietnamese operators, whose dimensions are all smaller than comparable U.S. measures (Kennedy 1972). Since difficulties in meeting control force requirements had been reported, the arrangement of controls was a major concern in this case. If the seat were kept as it was, difficulties in control operation occurred at the forward reach area (1), at the major manipulation area in front of the trunk above the thighs (2), and at the pedals (3).

For control operation in area 1, mainly small forces and torques were required, most of which could be applied while the operator leaned slightly forward. Consequently, no serious difficulties were expected here.

Control area 2 contained a major control, the stick. An evaluation of appropriate anthropometric dimensions showed that the stick control should be lower by about 3 cm and moved aft by about 7 or 8 cm for either Asian operator population. An assessment of the effects of such a relocation on the strength capabilities of the operators showed that it would not result in significant changes in strength under operational conditions. Hence, relocation of the control was deemed desirable but not mandatory.

Area 3 contained pedals that are essential for systems operation. Neither kind of Asian operator could reach the pedals sufficiently well to operate them throughout their full range. The situation is shown by the relatively small area of overlapping reach envelopes in Figure 10. Since it was sometimes necessary to apply 800 or 1,000 N to the pedals in their full forward position, relocation of the pedals by moving them backwards by 10 to 12 cm was mandatory.

SUMMARY AND CONCLUSIONS

Strength has been defined as the maximum force that muscles can exert in a single voluntary effort. Since strength cannot easily be assessed at the living muscle *in situ*, it is traditionally measured as the force or torque that can be exerted on some measuring device. The latter is, in fact, precisely the kind of information the human factors engineer needs for control design.

Both strength data and ergonomic design tasks are highly specific. There is no such thing as a "general strength" or a "universal design." Control parameters, like location, type, and directional requirements, relate to biomechanical factors which, in turn, determine the operator strength available under these specific conditions.

The strength exerted by a subject depends not only on his muscular capability but also on a number of experimental (technical, motivational, physiological, biomechanical and environmental) variables. Hence, standardization of strength-testing methods is highly desirable to provide reliable and comparable data needed for the design of tools, equipment, and complex man-machine systems.

Strength available for control operation is critical in setting either minimum force or torque levels, so that even the weak operator can manipulate the controls, or in selecting maximum limits to prevent accidental actuation or damage by too strong an operator. The designer selects an "optimum" value for control operation under normal conditions somewhere along the scale of minimum to maximum strength of the operator population. Percentile values and ranges are convenient for describing the sample data and for selecting design values.

Control location is also critical. The work space available for controls depends primarily on body dimensions and body positions. Within the total reach envelope of the hands and feet, certain locations allow very effective use of muscular strength, while other control locations are much less suitable.

Handedness can also be a decisive factor in control design. Lateral preference must be described in terms of the efficiency of performing tasks with either side of the body. Laterality interacts with expectancies about the way controls operate and the way systems respond to control movements. Such expectancies vary even between rather closely related populations. Anecdotal evidence suggests that larger stereotypical discrepancies exist between populations with more distinctively different histories and customs.

The biomechanical principles of the human body are the same for all populations. Hence, design for selected ranges of dimensions of joint angles, and of body positions applies to all user populations. In the past, design for such biomechanical principles has been difficult, mostly for rather practical reasons. However, new concepts of geometry, biomechanics, and ergonomics in computerized man-machine models promise to provide more efficient design tools for the ergonomist.

REFERENCES

Annett, M. The growth of manual preference and speed. *British Journal of Psychology*, 1970, **61**, 545–58.

Anthropology Branch. *Anthropometry of U.S. Air Force rated officers—1967.* Air Force Systems Command Handbook I-3. (1st ed.) Wright-Patterson Air Force Base, Ohio: Aeronautical Systems Division, 1969.

Astrand, P. O., and Rodahl, K. *Textbook of work physiology.* New York: McGraw-Hill, 1970.

Barnsley, R. H., and Rabinovitch, M. S. Handedness: proficiency versus stated preference. *Perceptual and Motor Skills,* 1970, **30**, 343–62.

Cathcart, E. P., Hughes, D. E. R., and Chalmers, J. G. *The physique of man in industry.* Report No. 71. London: Industrial Health Research Board, Medical Research Council, 1935.

Clarke, H. H. Analysis of physical fitness index test scores of air crew students at the close of a physical conditioning program. *Research Quarterly,* 1945, **16**, 192–95.

Crawford, W. A. Pilot foot loads. FPRC Memorandum number 57. Farnborough, Hants: RAF Institute of Aviation Medicine, 1954.

Daniels, G. S. The "average man"? Technical Note WCRD 53-7. Wright-Patterson Air Force Base, Ohio: Wright Air Development Center, 1952.

Elbel, E. R. Relationship between leg strength, leg endurance and other body measurements. *Journal of Applied Physiology,* 1949, **2**, 197–207.

Emanuel, I., Chaffee, J. W., and Wing, J. A study of human weight lifting capabilities for loading ammunition into the F-86H aircraft. Report number WADC-TR-56-367. Wright-Patterson Air Force Base, Ohio: Wright Air Development Center, 1956.

Faulkner, T. W., and Day, R. A. The maximum functional reach for the female operator. *American Institute of Industrial Engineers Transactions,* 1970, **2**, 126–31.

Garrett, J. W., and Kennedy, K. W. A collation of anthropometry, 2 volumes. Report number AMRL-TR-68-1. Wright-Patterson Air Force Base, Ohio: Aerospace Medical Research Laboratory, March 1971.

Gough, M. N., and Beard, A. P. Limitations of the pilot in applying forces to airplane controls. Technical Note 550. Washington, D.C.: National Advisory Committee for Aeronautics, 1936.

Guthrie, G. M., Brislin, R., and Sinaiko, H. W. Some aptitudes and abilities of Vietnamese technicians: implications for training. Paper P-659. Arlington, Virginia: Institute of Defense Analyses, 1970.

Hart, G. L., Rowland, G. E., and Malina, R. Anthropometric survey of the Armed Forces of the Republic of Korea. Technical Report Number 68-8-PR. Natick, Massachusetts: United States Army Natick Laboratories, 1967.

Hertel, H. Determination of the maximum control forces and attainable quickness in the operation of airplane controls. Technical Memo No. 583. Washington, D.C.: National Advisory Committee on Aeronautics, 1930.

Hertzberg, H. T. E. Dynamic anthropometry of working positions. *Human Factors,* 1960, **2**, 147–55.

Hertzberg, H. T. E., and Burke, F. E. Foot forces exerted at various brake-pedal angles. *Human Factors,* 1971, **13**, 445–56.

Hugh-Jones, P. The effect of limb position in seated subjects on their ability to utilize the maximum contractile force of the limb muscles. *Journal of Physiology,* 1947, **105**, 332–44.

Hunsicker, P. A., and Donnelly, R. J. Instruments to measure strength. *Research Quarterly*, 1955, **26**, 408–20.

Kennedy, K. W. Reach capability of the USAF population. Phase 1: The outer boundaries of grasping-reach envelopes for the shirt-sleeved, seated operator. Report number AMRL-TR-64-59. Wright-Patterson Air Force Base, Ohio: Aerospace Medical Research Laboratories, 1964.

Kennedy, K. W. International anthropometric variability and its effects on aircraft cockpit design. Report number AMRL-TR-72-45. Wright-Patterson Air Force Base, Ohio: Aerospace Medical Research Laboratory, 1972.

Kimura, D., and Vanderwolf, C. H. The relation between hand preference and the performance of individual finger movements by left and right hands. *Brain*, 1970, **93**, 767–74.

Kroemer, K. H. E. *Was man von Schaltern, Kurbeln und Pedalen wissen muss. Auswahl, Anordnung und Gebrauch von Betätigungsteilen*. Berlin: Beuth, 1967.

Kroemer, K. H. E. Human strength: terminology, measurement and interpretation of data. *Human Factors*, 1970, **12**, 297–313. (a)

Kroemer, K. H. E. Arbeitswissenschaft—ergonomie—human factors. *Werkstatts-technik*, 1970, **60**, 470–74. (b)

Kroemer, K. H. E. Foot operation of controls. *Ergonomics*, 1971, **14**, 333–61. (a)

Kroemer, K. H. E. Pedal operation by the seated operator. Report number AMRL-TR-71-102. Wright-Patterson Air Force Base, Ohio: Aerospace Medical Research Laboratory, 1971. (b) (Also published as SAE Paper 720004. New York: Society of Automotive Engineers, 1972.)

Kroemer, K. H. E. COMBIMAN-COMputerized BIomechanical MAN-model. Report number AMRL-TR-72-16. Wright-Patterson Air Force Base, Ohio: Aerospace Medical Research Laboratory, 1973. (Also published in *Proceedings of the IfU Colloquium: Space Technology-A Model for Safety Techniques and Accident Prevention. Cologne, Germany, April 1972*. Cologne: Verlag TUV Rheinland, 1973).

Kroemer, K. H. E. Designing for muscular strength of various populations. Report number AMRL-TR-72-46. Wright-Patterson Air Force Base, Ohio: Aerospace Medical Research Laboratory, 1974.

Laubach, L. L., Kroemer, K. H. E., and Thordsen, M. L. Relationships among isometric forces measured in aircraft control locations. *Aerospace Medicine*, 1972, **43**, 738–42.

Lehmann, G. (Editor), *Arbeitsphysiologie*. Volume 1 of *Handbuch der gesamten Arbeitsmedizin*. Berlin: Urban und Schwarzenberg, 1961.

Martin, W. B., and Johnson, E. E. An optimum range of seat positions as determined by exertion of pressure upon a foot pedal. Report No. 86. Fort Knox, Ky.: U.S. Army Medical Research Laboratory, 1952.

Morgan, C. T., Cook, J. S., Chapanis, A., and Lund, M. W. (Editors). *Human engineering guide to equipment design*. New York: McGraw–Hill, 1963.

Müller, E. A. Die günstigste Anordnung im Sitzen betätigter Fusshebel. *Arbeitsphysiologie*, 1936, **9**, 125–37.

Palmer, R. D. Development of a differentiated handedness. *Psychological Bulletin*, 1967, **62**, 257–72.

Pangle, R., and Garrett, L. Origin of the spring scale dynamometer. *Research Quarterly*, 1966, **37**, 155–56.

Rees, J. E., and Graham, N. E. The effect of backrest position on the push which can be exerted on an isometric footpedal. *Journal of Anatomy*, 1952, **86**, 310–19.

Reid, B. An annotated bibliography of United States Air Force applied physical anthropology. January 1946 to May 1973. Report number AMRL-TR-73-51. Wright-Patterson Air Force Base, Ohio: Aerospace Medical Research Laboratory, 1973.

Rohmert, W. Maximalkräfte von Männern im Bewegungsraum der Arme und Beine. Research Report number 1616 of the State of Northrhine-Westfalia. Cologne-Opladen: Westdeutscher Verlag, 1966.

Rohmert, W., and Jenik, P. Isometric muscular strength in women. In R. J. Shepard (Editor), *Frontiers of fitness*. Springfield, Illinois: Thomas, 1971, pp. 79–97.

Scherrer, J. *Physiologie de travail (ergonomie)*, 2 volumes. Paris: Masson et Cie, 1967.

Schmidt, R. T., and Toews, J. V. Grip strength as measured by the Jamar dynamometer. *Archives of Physical Medicine and Rehabilitation*, 1970, **51**, 321–27.

Snook, S. H., Irvine, C. H., and Bass, S. F. Maximum weights and work loads acceptable to male industrial workers. *American Industrial Hygiene Association Journal*, 1970, **31**, 579–86.

Switzer, S. A. Weight-lifting capabilities of a selected sample of human males. Report number MRL-TDR-62-57. Wright-Patterson Air Force Base, Ohio: Behavioral Sciences Laboratory, 1962.

Van Cott, H. P., and Kincade, R. G. (Editors), *Human engineering guide to equipment design* (revised edition). Washington, D.C.: U.S. Government Printing Office, 1972.

White, R. M. Anthropometric survey of the Royal Thai Armed Forces. Natick, Massachusetts: United States Army Natick Laboratories, 1964. (a)

White, R. M. Anthropometric survey of the Armed Forces of the Republic of Vietnam. Natick, Massachusetts: United States Army Natick Laboratories, 1964. (b)

Whitley, J. D., and Allen, L. G. Specificity versus generality in static strength performance: review of the literature. *Archives of Physical Medicine and Rehabilitation*, 1971, **52**, 371–75.

SIX

The Role of Human Factors Engineering in Underdeveloped Countries, with Special Reference to India

CHITTRANJAN N. DAFTUAR

It is not easy to classify unequivocally a country as developed or underdeveloped. According to Pepelasis, Mears, and Adelman (1961) some underdeveloped countries have the material requisites, human potentiality, and willingness to make economic progress, but they suffer from mismanagement of their resources. Pepelasis et al. argue further that, given proper opportunities and incentives, most of these countries have the potentialities of becoming "achieving societies." This point of view is a good starting point for human factors specialists in developing countries.

To achieve conditions requisite to industrialization and social change, individual and social learning has to be fostered through carefully planned programs of research and development. Alexander (1962) has identified the following seven conditions that must be remedied for economic growth: (1) low per capita income, (2) an unbalanced economy, (3) untapped natural resources, (4) a tradition-oriented culture, (5) a large but untrained labor force, (6) a small amount of capital equipment, and (7) chronic underemployment. To meet such a challenge, McClelland (1961) has suggested a series of psychological objectives:

- The gradual substitution of such conflicting values as "inner directedness" or "caste," with "other directedness" and "market morality."
- The substitution of father-figure dominance by habits of independent choice and action.

Chittranjan N. Daftuar, Department of Psychology, Gaya College, Gaya, Bihar, India.

- The introduction of ideological reforms to unify other directedness, market morality, and decreased father-dominance.
- The gradual introduction of educational programs with both short- and long-term benefits in the basic skill and knowledge requirements of a technological society.
- The reorganization of fantasy life to conform with a new cultural milieu.
- The more efficient use of existing need achievement resources.

To sum up, the "achieving society" must undertake broad, society-wide programs if it is to industrialize successfully. At the same time, one must be mindful of the differences between the psychological strategies needed for cultural change and the tools needed to create such changes. Programs of financial aid, education, improved roads and transportation systems, and increased communications among the members of a culture are only tools to be used in the fulfillment of strategic objectives. These tools must be coupled with a rational, planned program based on human engineering principles that should in turn be based on the local requirements, habit system, and psychological and anthropometric limitations of a given population.

PART 1: SOME DISTINCTIVE INDIAN CULTURAL AND ECONOMIC PROBLEMS

India is a country with mixed levels of technology. On the one hand, it constructs and maintains nuclear reactors and sophisticated electronic equipment for both military and civilian uses. On the other hand, possession of an automobile is a status symbol to 80 percent of its population. Likewise, since agricultural methods have only started to change, a tractor in rural India, like a car in the cities, is a matter of considerable status. Such extreme diversity has also to be viewed against a background of unique social, human, and behavioral peculiarities.

THE FAMILY AND THE CASTES

Rural India is a closely knit society with commonly shared work and family responsibilities operating through two important social institutions: the family system and the castes. Although there are indications lately that the family is showing some signs of disintegration, it, and not the individual, is still the basic social unit (Datta 1961). Village life is still tradition-bound, and jobs in villages are generally allocated and designed on the basis of family and caste. The son of a carpenter is most likely to become a carpenter, and he will generally inherit from his father not only his job, but his tools and postural patterns. This probably explains why attitudes about regular attendance at work to better oneself and family and a competitive philosophy of life are lacking in this country. In traditional India a job is generally regarded as a family responsibility to such an extent that its performance is shared by all the members of a family. For example, if the father is a blacksmith he will generally be assisted by his son in such auxiliary tasks as assembling and the transport of tools. Moreover, attempts to bring in anyone from outside the family are likely to be resented.

This kind of thinking was responsible for a widespread resistance movement in the Tata Iron and Steel Company (TISCO), Jamshedpur, some time ago. In the end, management acceded to worker demands that in filling all future vacancies the company would absorb at least one descendant of currently employed workers. This practice is still observed in TISCO and certain other companies.

To take another instance, Fraser (1966) states that the failure of a weavers' cooperative started at Barpali Village Service was due in part to the fact that the project technicians, interested only in selecting the best workers, had drawn weavers from two different caste groups that could not by tradition work together.

AGRICULTURE

Indian agriculture is a mixture of the old and the new. Although some farmers use modern tractors, most farmers still use tools that are essentially unchanged from those of a thousand years ago: homemade yokes for oxen, rope from local plants, and iron plowpoints and axeheads fashioned by the village blacksmith. Threshing is often done by oxen walking round and round over stalks of cereal until the grain is loosened sufficiently for winnowing. Alternatively, cereal stalks are threshed by human hands. For the latter purpose a rectangular wooden platform made out of tree trunks is used. The wooden platform is kept on a wooden bed and the grain is threshed by whipping the stalks of cereal on the flat wooden platform (see Fig. 1). This description conveys a significant message: For Indian agriculture we need to consider not only man-machine but man-machine-animal systems.

In rural India, time and labor are abundant resources, while capital to buy modern tools is extremely limited. These realities force Indian ergonomists to concentrate on small tools, with economy as a primary consideration.

HOUSING AND ARCHITECTURE

Thatched huts with mud walls, constructed usually by a family, its relatives, and neighbors, are still the most common kind of shelter in villages. The hutments are generally open to the elements and this, together with the limited kinds of clothing and other protective materials used by the inhabitants, poses some real problems for both building designers and ergonomists (Daftuar 1971*a*). Recently the federal government initiated an

Fig. 1. Threshing by hand in rural India.

extensive program of low-cost housing in rural areas. It has yet to be seen how far this program will solve its human factors problems. One thing is clear: Innovators who bring their own traditions with them may fail to make use of locally available materials and skills and may be responsible for the introduction of housing unsuited to local conditions. India has a hot and humid climate. In cities there is a tendency, due largely to the British legacy, to use western structural and design techniques—heavy bricks with minimal ventilation. Such houses are not only very damp and uncomfortable, but their construction requires skills that are usually not found in local village populations.

CLOTHING

Indian workers, in villages as well as in cities, generally work half clad, exposing their bodies to severe wind and chill. The problem is intensified because a great deal of work in India involves handling materials that are heavy, cold, wet, and rough. Inadequate clothing is a serious problem in mountainous regions, especially for the armed forces. India has a very long border with the Peoples Republic of China. Parts of this border are in the

Himalayas. It is generally recognized that one major reason for the Indian debacle in 1962 was the poor clothing provided for the armed forces who often had to fight in below freezing temperatures.

HOUSEHOLD UTENSILS

The utensils and pottery used by Indian housewives in villages are made of clay. They are usually made by members of the village pottery caste with the help of wheels. Rotation of these wheels is started by a motion of the hands and momentum is imparted to the wheel by a stick held in the hands. The work of pushing the wheels may require a considerable amount of human energy. Fortunately, this method is gradually being replaced due to the increased use of metal utensils in villages as well as in cities.

WORKING POSTURES

Working postures differ among different cultural groups (Tichauer 1963). Indian workers more often squat than stand or sit (Daftuar & Bhan 1966). Figure 2 illustrates some typical working postures. For example, a carpenter working in a saw mill will most probably squat on the ground rather than assume any other working posture. Similarly, a housewife usually prefers to squat on the ground in the preparation of family meals, even if she possesses such modern kitchen appliances as a gas or electric oven. For this reason Ambee Industries (Ahmedabad, India) has designed a gas oven that can be easily manipulated on the ground or, at most, on a platform 12 or 13 cm above the ground. A full description of this stove is given in Part 3 of this paper.

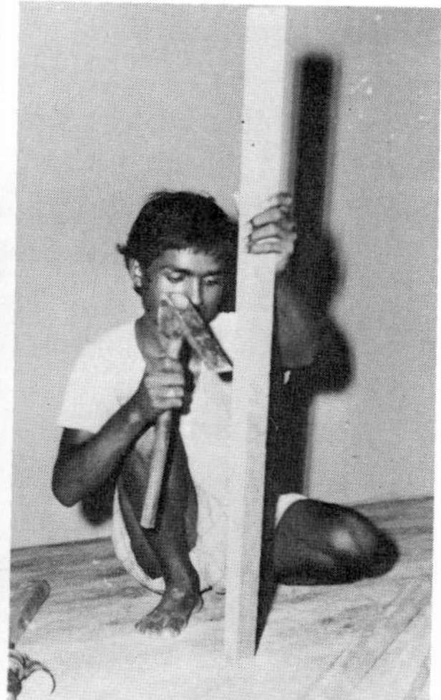

Fig. 2. Typical Indian working postures assumed by a housewife preparing food (*left*) and a carpenter working on wood (*right*).

CONCEPT OF EFFICIENCY

The concept of *efficiency* in Indian thinking is different from what is understood in the West. To refer again to the Barpali Village Service Project, McClelland and Winter (1969) concluded that another reason for the failure of the project was that the norm of efficiency conflicted with caste traditionalism. Raising poultry is a low-caste occupation, and those who engaged in it did not want to improve their performance lest it prevent them from rising socially.

The extended family often forces an enterprising man to spend whatever he has saved on supporting others rather than on investment in business expansion. In traditional India the efficiency of a man is judged not by economic achievements alone but also by the effectiveness with which he maintains his extended family. As a result, job satisfaction in India is generally more directly related to a worker's satisfaction with his home environment than it is in the case of his counterparts in the West (Daftuar 1969).

ANTICIPATING INNOVATION

Another area where human factors engineers can help is in educating and training the masses for innovation. A well-designed learning program can foster better adjustment to rapid technological change. Once again, such programs must be appropriate for local conditions. In India, for example, visual aids should generally not show too advanced technology nor require sophisticated powers of interpretation. Nor should such aids violate Indian customs. A film successful in some areas may fail in India simply because a woman is shown in deep red or in some other very glossy colored dress. In India such dresses are usually worn by brides and by other women on special occasions, such as festivals. A simple peasant woman may be confused by what a festival dress has to do with the purpose of the film. For a country like India it is also essential that as much training be devoted to the maintenance of tools and machinery as to the introduction of technological innovations. Many foreign-made tractors are soon inoperative in Indian villages, because nobody has been properly taught to maintain them. The villagers are trained only to use the tractors.

AUTOMATION

While large-scale automation is increasing rapidly in almost every field in the West, the installation of high-speed computers has become a real problem for various Indian managements. Designing automated jobs in India is not an easy task. A case study (personal communication from U.S. Prasad 1972) of the Life Insurance Corporation of India (LIC) illustrates some special problems peculiar to the Indian scene.

In 1964 the All India Insurance Working Committee discussed the reorganization of LIC's working system and recommended the installation of high-speed computers. The first high-speed computer was installed in

LIC's Bombay Zonal Division in September 1965. That installation resulted in a large-scale strike and a resistance movement among LIC workers across the nation. The strikes were complete and nation-wide. A poll in August 1968 showed that 97 percent of LIC workers favored a general and indefinite strike. Subsequently, an *All India Committee Against Automation* and several other related committees were formed. Five million signatures were collected against automation, and these were submitted to the Indian Parliament. Almost all the political parties, including the then ruling party, and the trade unions of the country supported the movement. In Calcutta, the Ilaco building was virtually in a state of siege for three to four months during 1967–68. Dock workers were not even allowed to unload computer parts. The United Front government (a mixed party government) of West Bengal also refused to allow police to help keep the machine in the building.

The workers' main reason for their resistance was that, according to their estimate, 30,000 employees out of the then existing strength of 40,000 employees were likely to be laid off as surplus or redundant if the computer was put into operation. The government failed to give any assurance against this charge. The workers further argued that both their service security and their chances of promotion were in danger, since there would be no further appointments and hence no expansion of manpower in the corporation. Finally, they argued that high-speed computerization would hamper the servicing of policy-holder accounts because everything would be centralized in the four zonal divisions of Calcutta, New Delhi, Bombay, and Madras.

The crux of the entire case study for our purposes is that while LIC workers still oppose high-speed computerization, the government has gone ahead with the installation of small and medium computers in almost all divisional headquarters of the LIC. Nobody seems to be concerned about them. The lesson appears to be that if you want to design an automated job in India you must start with small or medium automation plans. Indian workers will apparently not object to that. But if the automation is large-scale, involving a very high-speed computer, you are heading for high-speed resistance.

PART 2: SOME STUDIES OF VISUAL DISPLAYS

Now let us turn to summaries of some hitherto unpublished studies of visual displays differing from similar ones published in the West because they involve customs or materials that are distinctively Indian.

THE MEANINGFULNESS OF INDIAN ROAD SIGNS

Indian road signs are, in general, old British road signs. Although the United Kingdom in the meantime switched to international road signs many years ago, Indian road signs have remained unchanged. Indeed,

Indian designers have paid no attention to the problem. The purpose of this study was to assess the meaningfulness of road signs currently used throughout India.

PROCEDURE

Tests were made on forty-five undergraduate psychology students with no driving experience. All the students had normal visual acuity and normal color vision as measured by standardized tests. Their ages ranged from sixteen to twenty-five years.

Out of a total set of 39 Indian road signs (Public Vehicle Department 1962), 25, those illustrated in Figure 3, were used. The remaining 14 were too obvious, or rare, and were dropped. The signs were each photographed in black and white in an actual road setting (see, for example, Fig. 4). The photographs were individually printed on 10.5 × 15.5 cm cards.

The students were tested individually. When a student came into the laboratory, he was first given standardized instructions. Then he was handed the photographs, one by one, and asked to interpret the meaning of the sign shown and to write his interpretation on a specially prepared answer sheet. Each photograph was presented to a subject for one minute and the subject was asked to scan the entire scene while concentrating on the traffic sign. The total experimental session for each subject lasted about 1½ hours and included a 15-minute rest period after the completion of the thirteenth photograph.

RESULTS

Since our sample consisted of students who did not drive, the percentages of correct interpretations of our road signs were naturally smaller than those obtained in comparable studies in the West. In fact, correct interpretations of the Indian road signs varied from zero to 75.5 percent with a mean of 39.9 percent (Table 1). Sign 24 (Fig. 3) was not understood by any subject, and no single traffic sign was correctly understood by all the subjects.

Some road signs were even found to convey meanings opposite to those actually intended. For example, the sign for *No Horn* (17 in Fig. 3) was often interpreted as *Blow your horn please*. The sign for *No overtaking* (16 in Fig. 3) was seldom interpreted as such. This was also the case with the sign for *Road crossing ahead* (24 in Fig. 3).

DISCUSSION

Symbolic displays may contain very little information and may be misinterpreted largely because of their poor design (Brown 1968). In a survey carried out in England, Mackie (1966) found that some symbolic displays were correctly interpreted by as few as 16 percent of drivers. Due to the nature of our sample, we expected still lower percentages. However, the fact that some signs were not correctly interpreted by a single subject calls

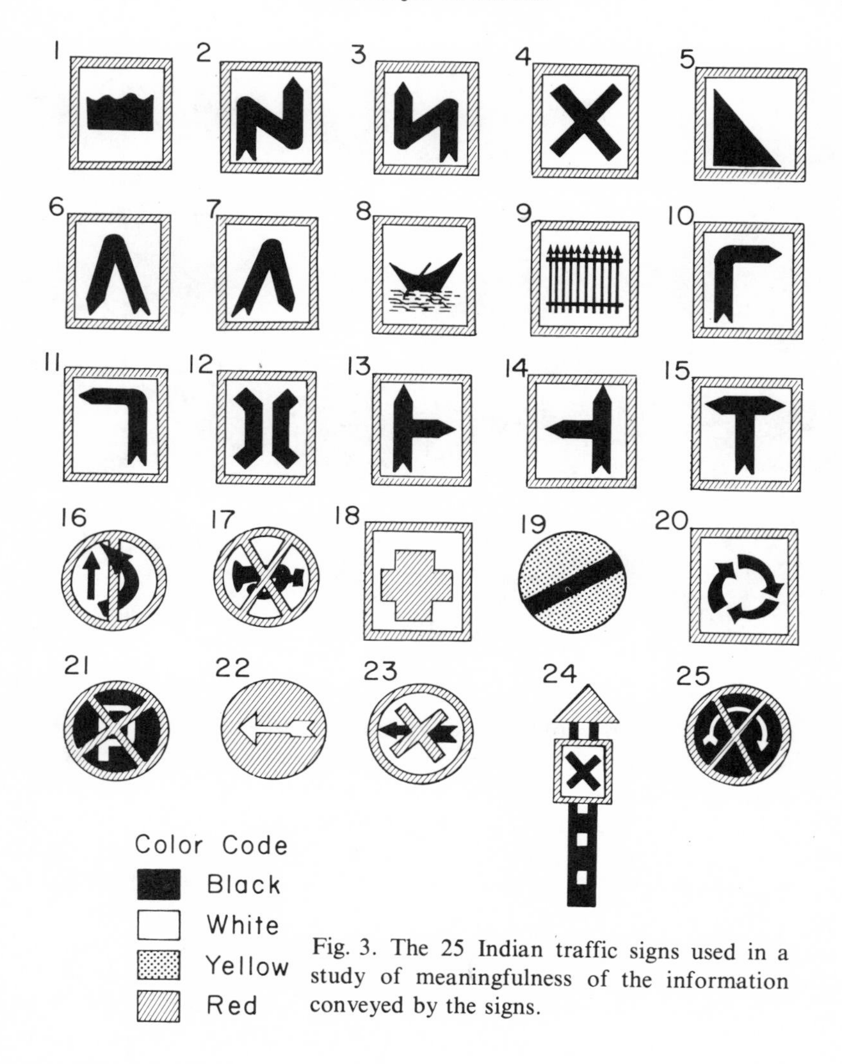

Fig. 3. The 25 Indian traffic signs used in a study of meaningfulness of the information conveyed by the signs.

for an urgent revision in the design of existing Indian road signs. For untrained persons they suffer from serious communication gaps. Since drivers are rarely tested for their knowledge of traffic signs at the time they are given driving licenses in India, the present results can probably be generalized to newly licensed drivers as well.

THE RELATIVE LEGIBILITY OF ALPHANUMERIC CHARACTERS OF ROMAN, ARABIC, DEVENAGARI, AND BENGALI SCRIPTS

India has several regional languages, hundreds of dialects, and many scripts. Under the Indian constitution, Hindi was envisioned as the Indian

Fig. 4. Each road sign in Figure 3 was shown in an actual setting such as this one.

national language because it is common in almost half of the nation. Following independence, the Indian leadership decided to switch over gradually to Hindi. However, when the time came for the switch, a storm of controversy arose in non-Hindi-speaking areas. As a consequence, English has remained the official language of the federal government, while the various regional languages remain as the official languages of their respective states. In the 1960s, there was a widespread movement in the Hindi-speaking areas for the introduction of Hindi at all levels of education and throughout the government. Hindi fanatics went so far as to paint over all English signboards and nameplates in black. Even though Arabic numerals developed originally from the Devenagari script, these fanatics even painted over automobile license plates because they bore Arabic numerals.

Various suggestions have been made for an amicable resolution of these controversies. One suggestion is to use Roman script for all the various Indian languages. Throughout all of these controversies, proposals, and counterproposals, no one has inquired into the scientific merits of these several languages for the practical communication of information. This study was designed to test first the relative legibility of Arabic, Devenagari, and Roman numerals, and second the relative legibility of two Indian scripts, Hindi, spoken by the largest portion of the Indian population and the official Indian national language, and Bengali, having the richest litera-

Table 1. Numbers and percentages of 45 students who interpreted correctly the road signs in Figure 3

Number of road sign	Number correct	Percentage correct	Number of road sign	Number correct	Percentage correct
1	12	26.7	14	22	48.9
2	8	17.8	15	4	8.9
3	6	13.3	16	1	2.2
4	17	37.8	17	17	37.8
5	13	28.9	18	27	60.0
6	12	26.7	19	22	48.9
7	13	28.9	20	11	24.4
8	19	42.2	21	14	31.1
9	34	75.5	22	10	22.2
10	26	57.8	23	9	20.0
11	24	53.4	24	0	0.0
12	28	62.2	25	9	20.0
13	19	42.2			

ture among all the Indian languages; and one foreign script, Roman, still used by the intelligentsia and bureaucracy almost all over India.

LEGIBILITY OF NUMERALS

Although Arabic numerals are a foreign script, they are almost a second mother-script for the Indian populace. Even barely literate persons can read and write Arabic numerals as well as their counterparts in Devenagari. Since there was no previous literature to guide us for formulating any specific hypothesis, we expected only that, since we used subjects having Bengali as their mother tongue, the alphabet of Bengali scripts would be most legible for our subjects.

Subjects. Twenty undergraduate and postgraduate students acted as subjects. All belonged to the Bengali-speaking community in Bihar, a Hindi-speaking area. We ensured that all of them knew and were equally familiar with the Roman, Devenagari, and Bengali scripts. They had a mean age of 22.6 years with a range of 18 to 30 years. There were equal numbers of men and women, and they were selected equally from the arts and sciences. All the subjects had normal vision. Corrected vision was not permitted.

Apparatus and test materials. The stimuli were presented with an ordinary tachistoscope. Experiments were conducted in a laboratory and under daylight conditions. Stimuli were printed on white cards in deep black printing ink (see Fig. 5). Three sizes—8-, 10-, and 12-point—were used for all three kinds of numeral. The digits 645, 768, 794, 684, and 975 were reproduced in all three sizes and scripts, yielding a total of 45 stimuli: 5 sets of digits × 3 scripts × 3 sizes. The digits 1, 2, and 3 were not tested because they are exactly alike in Arabic and Devenagari.

Experimental procedure. Each subject sat so that there was a constant distance of 76.2 cm between the stimuli and his cornea. This adjustment was maintained with the help of a chin rest. Stimuli were exposed for one-fifteenth of a second.

Tables of random numbers were used to assign a random order to all forty-five cards for the purpose of presentation. This random order was used for half the subjects and the reverse order for the other half.

To get the subjects fully accustomed to the experimental procedures they were given practice trials to the criterion of one correct response. Standardized instructions about the procedures and purpose of the experiment were given.

Subjects were asked to pronounce the digits they saw and their responses were recorded by an assistant as either correct or incorrect. To be called correct all three digits had to be correctly read. Thus, for any one block of five stimuli, the maximum score for a subject was 5.

After every fifteen trials, subjects were given a rest period of 5 minutes. An entire experimental session for one subject lasted about an hour.

Results. An analysis of variance of the data showed that there were highly significant differences among scripts, sizes of type, and the combinations of script and size, that is, the interaction of script with size. The largest single effect, however, was attributable to script (see Fig. 6). The Roman digits were the least legible of those tested. Although the average legibility score for the Arabic numerals was less than for the Devenagari numerals, the difference is small and not statistically significant. Legibility scores tend to increase as the size of type increases, but the exceptions to this generalization are sufficiently large to make a statistically significant interaction. I have no ready explanation for these irregularities in the data.

LEGIBILITY OF THE ROMAN, DEVENAGARI, AND BENGALI ALPHABETS

Subjects. The subjects who participated in this study were the same as those who had participated in the study of numerals.

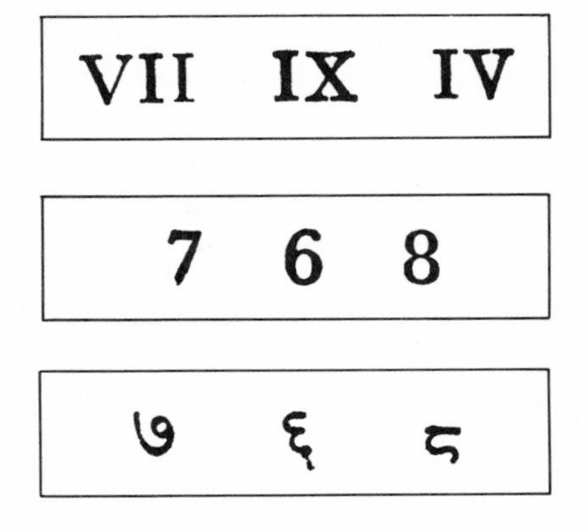

Fig. 5. Example of the Roman (*top*), Arabic (*center*), and Devenagari (*bottom*) numerals. (*Editor's Note*: No attempt has been made to make the sizes of the numerals in this illustration agree with any of the sizes of the numerals tested.)

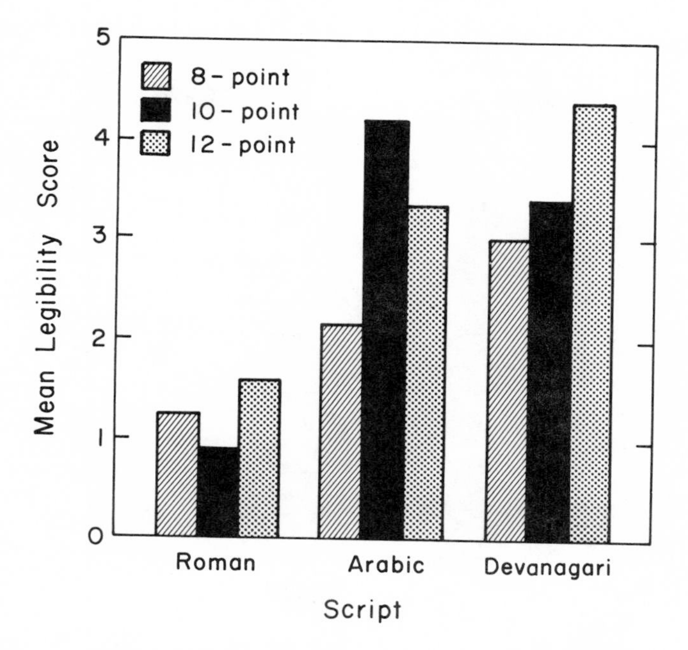

Fig. 6. Mean legibility scores for three types of numerals.

Apparatus and test materials. The same tachistoscope and the same experimental procedure and design were used in this study as in the numeral study.

The stimuli for this experiment were selected on the basis of results from a preliminary experiment to identify CVCVC paralogs of zero association value in Devenagari (Daftuar 1972). Similar five-letter, CVCVC paralogs were selected for the Bengali and Roman alphabets. Ten paralogs of each of the three scripts were used as stimuli. Four point sizes—8-, 10-, 12- and 14-point—were to have been used. However, the present analysis is based on the results obtained with 12- and 14-point printing faces only, because the 8- and 10-point faces for the Bengali scripts were not available in any local press.

We used CVCVC paralogs for two main reasons: (1) to minimize the possibility of bias in favor of any language; and (2) because Hodge (1963) has suggested that five-letter stimulus materials are capable of giving more realistic data than single letters.

The stimuli were printed on white cards in deep black ink. In the case of the Roman alphabet, only upper-case letters were used. Devenagari and Bengali scripts have only one case. Sample cards are shown in Figure 7. A stimulus set consisted of 60 cards, 10 paralogs × 3 scripts × 2 sizes.

Fig. 7. Examples of the Roman (*top*), Devenagari (*center*), and Bengali (*bottom*) scripts. (*Editor's Note*: No attempt has been made to make the size of the letters in this illustration agree with either of the sizes of the scripts tested.)

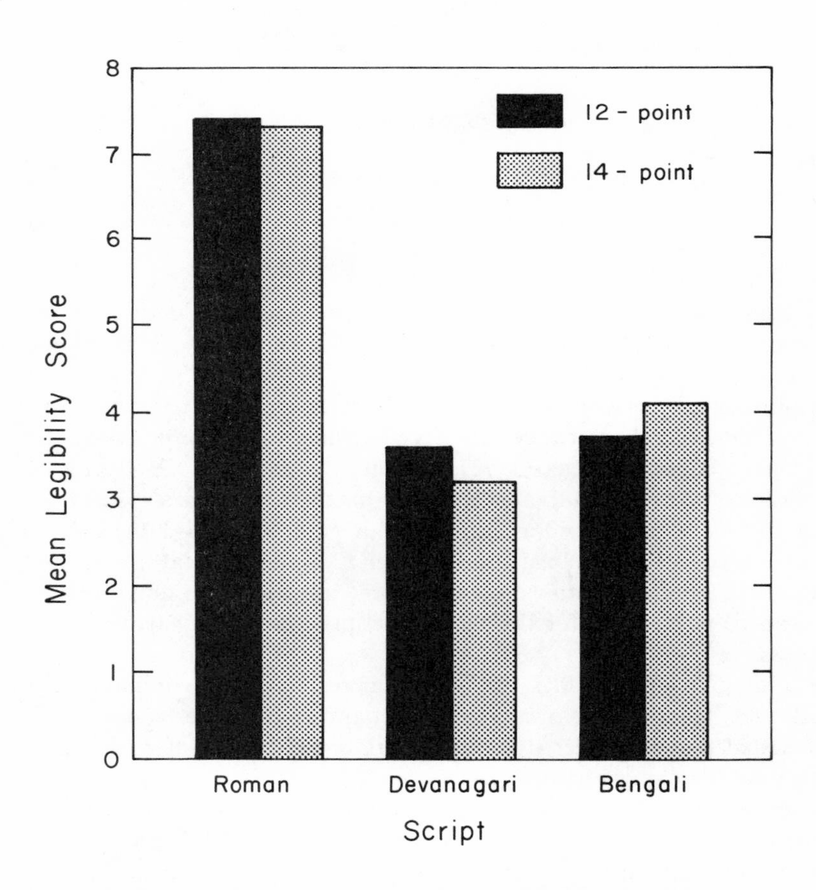

Fig. 8. Mean legibility scores for three types of alphabet.

Experimental procedure. The experimental procedure for this study was exactly the same as that for the study on numerals. Subjects were asked to pronounce what they saw, and their responses were recorded by an assistant. The responses were later scored in the same manner as for the numerals. Thus for any one block of stimuli, comprising ten stimuli in each case, the maximum score for a subject was 10.

After every ten trials subjects were given a rest period of 5 minutes. The entire experimental session lasted about 1½ hr. Two weeks elapsed between the experiment with the numerals and this experiment.

Results. An analysis of variance on the data shows a highly significant difference among scripts. Neither the difference between sizes nor the interaction of scripts with size was significant. The mean data (Fig. 8) show that the Roman script is distinctly more legible than the Devenagari and Bengali scripts, but that the latter two do not differ appreciably from each other.

DISCUSSION

In the absence of precise hypotheses we had expected the Bengali alphabet to be most legible for our subjects. This was clearly not the case. The Roman alphabet was far superior to either Indian alphabet. One likely explanation is that Roman letters are far less confusing in their structure. Out of a total of twenty-six upper-case Roman letters, sixteen are constructed by joining simple horizontal, vertical, or perpendicular lines. In the Devenagari or Bengali scripts, by contrast, there are hardly three or four such structures.

Empirical evidence to support this explanation comes from a comparison of results reported by Hodge (1962) and Howell and Kraft (1959), on the one hand, with those reported by Aziz (1970) on the other. Hodge found very low percentages of confusion among the Roman letters. Similarly, Howell and Craft report a maximum of about twenty percent confusion among Roman letters. By contrast Aziz (1970) reported confusion scores as high as 25.6 percent for the Bengali alphabet under normal laboratory conditions. Since there is a considerable amount of structural similarity between Devenagari and Bengali, it can be assumed that Devenagari would probably yield similar confusion data. Research is needed to test this assumption.

Although these findings are only suggestive, they point to what appear to be some fundamental differences among Indian and Western scripts. This is an exciting area of research that clearly needs to be extended. It may not only reveal important differences in legibility among the various alphabets and numerals of the world, but may also uncover some basic secrets about the workings of human information processing. In any case, much more work of this kind has to be done before we can make truly universal human engineering recommendations about alphanumeric symbols to be used in all cultures and all nations.

PART 3: A DIGEST OF SOME HUMAN FACTORS WORK IN INDIA

The literature on human factors work in India is meager. Readers interested in that literature are referred to reviews by Baumgartel (1966), Daftuar (1969, 1971*b*), and Sinha (1970, 1972). My intention is not to review comprehensively the scanty material available, but rather to mention briefly a few studies that bear directly on the topic of this book. These are studies that are either (a) cross-cultural, that is, that involve Indian subjects and those of some other nationalities, or (b) are concerned with problems that are distinctly Indian in nature. The studies mentioned here are also not generally available in the West.

A STUDY OF FORESTRY WORKERS

A truly cross-cultural field study was conducted under the cooperation of the Institute of Work Physiology, Stockholm, and the Forest Research Institute, Dehradun (Hansson, Lindholm, & Birath 1966) through funds made available by the government of Sweden under an Agreement on Financial Development Cooperation (1964/65). The primary purpose of the study, which lasted for seven weeks, was to compare different saws used by trained forestry workers. Although the study produced a great many results, only the following are relevant to our purposes:

1. Measurements were made of the body weight, height, and work capacity of forest workers from three regions of India and of Swedish and Norwegian forest workers. I have compared these data with similar measurements of Indian textile workers (Sen Gupta & Sen 1964) and coal miners (Chakravarty & Guharay 1965). The Indian workers had body weights ranging from about 46 to 51 kg, as compared with an average of about 72 kg for the Scandinavian workers. The Indian forest workers were also 11 to 13 cm shorter than their Scandinavian counterparts. One conclusion of the study was that logging tools with large physical dimensions are impractical in India.

2. In continuous work, the maximum O_2 consumption per kg of body weight was almost the same for the two samples. However, the maximal aerobic work capacity, expressed in liters of O_2 per min, of the Indian workers was 57 to 67 percent that of the Scandinavians.

3. The Indian system of felling by saw, lopping standing trees, turning logs, and loading timber imposes heavy local loads on the back muscles and hence unnecessary risks of injury.

4. The physiological load was high and the work output was low, with the one-man crosscut saw in felling. This is a one-man task, and sawing with a one-man crosscut saw requires a standing or kneeling position. Considering the work capacity of the Indian forest workers, the relatively harder wood, and the generally larger dimensions of the trees, all of which make it necessary for a man to work in a difficult position for a long time,

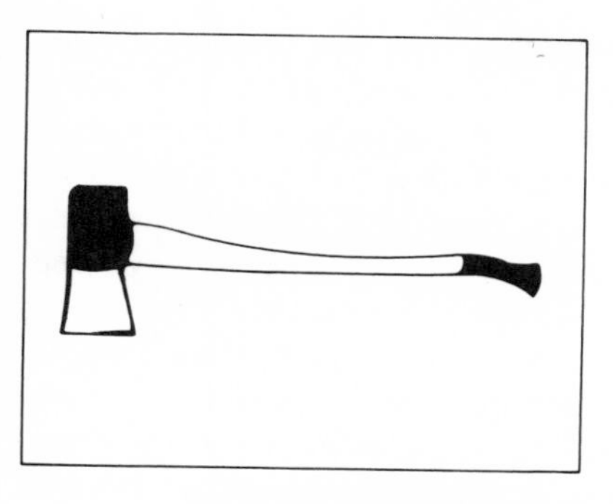

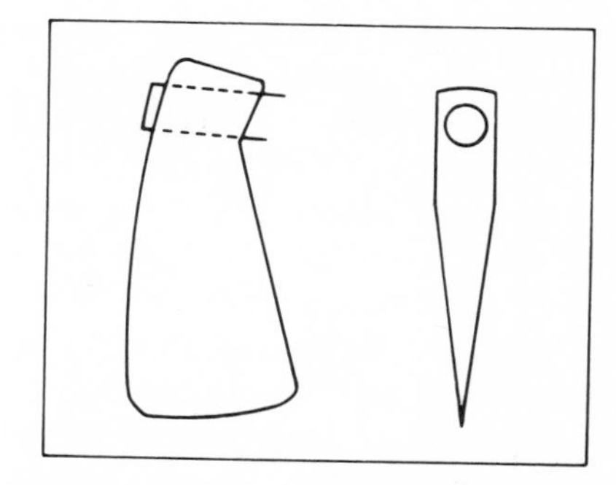

Fig. 9. Swedish (*left*) and Indian (*right*) axes (Hansson, Lindholm, & Birath 1966).

the one-man crosscut saw was not regarded as suitable for Indian conditions.

5. A Swedish axe, the "Saterpilen," and local Indian axes were tested in barking, lopping, and undercutting (Fig. 9). The Swedish type of axe was better than the Indian axe in undercutting and lopping, but for barking the Indian axe proved to be better. This might be because the Swedish axes had sharper and thinner edges, were made of better steel, and had a better balance and shaft. On the other hand, the Swedish axes had too small a back in proportion to the power developed and a bent shaft is unsuitable if the back as well as the edge of the axe is to be used. These drawbacks make the Swedish axe less useful in barking operations.

INDIAN WORKING POSTURES

As mentioned earlier, Indians prefer to work in a squatting posture. Dhesi and Firebaugh (1972*b*) measured the angular positions of various body members in the normal sitting position and in the making of *chapati*, Indian flat bread. Spots of adhesive were placed on various landmarks on the subject's body, the subject was photographed, and tracings were made from the photographs (see Fig. 10). Figure 10 defines the various angular measurements given in Table 2. So, for example, the mean angular deviation of the upper back from the vertical was 35.3°. This increased to 42.4° during the rolling of *chapati*. The data in Table 2 are difficult to relate to anthropometric data in Western human engineering guides, because the working posture illustrated in Figure 10 does not occur in the West.

SOME PHYSIOLOGICAL IMPLICATIONS OF THE SQUATTING POSTURE

In another study, measuring the effect of body position on heart rate, Dhesi and Firebaugh (1972*a*) concluded that significant changes in heart rate occurred in two out of three task conditions for subjects in the squatting position. The greatest increase in heart rate occurred during the rolling stage of the task. In this position, the subject is required to lean forward with a resultant compression of the abdominal and pelvic organs. Heart rate was also high during the puffing stage, when the subject sits with his body twisted. The partial static contraction of the muscles of the left sternocleidomastoid and deltoid muscles apparently interferes with the circulation of the blood. Of the seven different parts of the body measured, only the positions of the knee and ankle were significant in producing changes in heart rate. Comparing these findings with those of Hanson and Jones (1970), one may conclude that the Indian working posture, squatting, is perhaps a better posture than the customary Western sitting posture. Hanson and Jones tested subjects seated in various postures on a stool and in the squatting position. Although the heart rate was less in the squatting position than in the seated position, their subjects found the squatting posture difficult to maintain.

DESIGN OF A MANUALLY OPERATED HARVESTER

Saran and Ojha (1967) reported the redesign of a harvester to utilize man more efficiently and to match the financial and sociological limitations of rural India (Fig. 11). The harvester is designed for economy, versatility, and simplicity. As the operator pushes the harvester, power is transmitted from the traction wheels to the camshaft through a chain and sprocket assembly. The camshaft drives the cam, which actuates the fol-

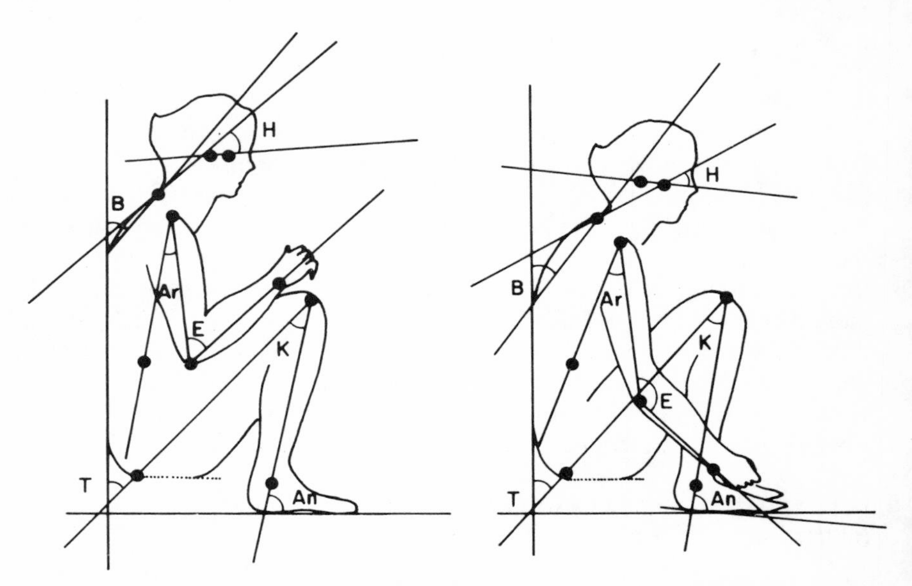

Fig. 10. Tracing of the normal sitting posture of a housewife, showing the various landmarks and angular measurements made (*left*). Outline tracing of the same housewife during the making of *chapati* (*right*) (Dhesi & Firebaugh 1972*b*).

Table 2. Mean angular positions of various parts of the body in the normal sitting position and during ball-making, rolling, and puffing in the making of *chapati* [a]

Positions of body during	Part of the body						
	Upper back (B)	Head tilt and thrust (H)	Armpit (Ar)	Elbow (E)	Thigh (T)	Knee (K)	Ankle (An)
Normal sitting	35.3	36.8	30.7	119.0	44.7	33.6	81.3
Ball-making	39.6[b]	37.5	48.4[c]	87.2[c]	41.6[c]	33.2	85.2[c]
Rolling	42.4[c]	33.7	61.0[c]	143.7[c]	42.9[b]	32.8	85.2[c]
Puffing	37.4[b]	57.2[c]	19.3[c]	94.2[c]	41.5[c]	34.2	85.8[c]

[a] Dhesi & Firebaugh 1972*b*.
[b] Significantly different from the normal sitting position at $p \leqslant 0.01$.
[c] Significantly different from the normal sitting position at $p \leqslant 0.001$.

lowers to give a reciprocating motion to the knife. A grass bar pushes the severed stalks to the side. This redesign utilizes human power more efficiently. The human energy requirements are well within the limits of the average Indian farmer, while the working speed and the height of the handle match his physical characteristics.

A KITCHEN STOVE FOR INDIAN HOMES

As mentioned earlier, Ambee Industries (Ahmedabad) has designed a gas oven so that it can be easily manipulated by Indian housewives accustomed to preparing food on the ground in a squatting position. The total length of the gas oven is only 58 cm, the total height, including the flame area, is 13 cm (10.5 cm for the oven and 2.5 cm for the flame area), the width at both ends is 29.5 cm. It has two flame areas, each covering 21 × 21 cm, that is, 441 square cm.

A KITCHEN ADAPTED TO THE INDIAN STYLE OF LIFE

Sinha (N.C.P. Sinha, personal communication, April 2, 1972) has designed a kitchen for his house which fits very well the anthropometric dimensions and the social and economic limitations of a typical Indian middle-class family (see Fig. 12). The entire kitchen area is 4.1 m in length, 1.8 m wide, and 2.3 m high. These dimensions easily fit even the 95th percentile of the Indian female population (Daftuar 1964; Chatterjee & Daftuar 1966). Within this small area the kitchen has a water tub, a three-story rack, storage space for coal and cow-dung fuel, three *Chulhahs* (ovens), a water tap, a small place to wash utensils, and a 30.5 cm high elevation to keep utensils. The kitchen designed by Sinha appears to combine a utilitarian and functional approach. The rising population in India and the increasing migration to cities have posed serious problems of ac-

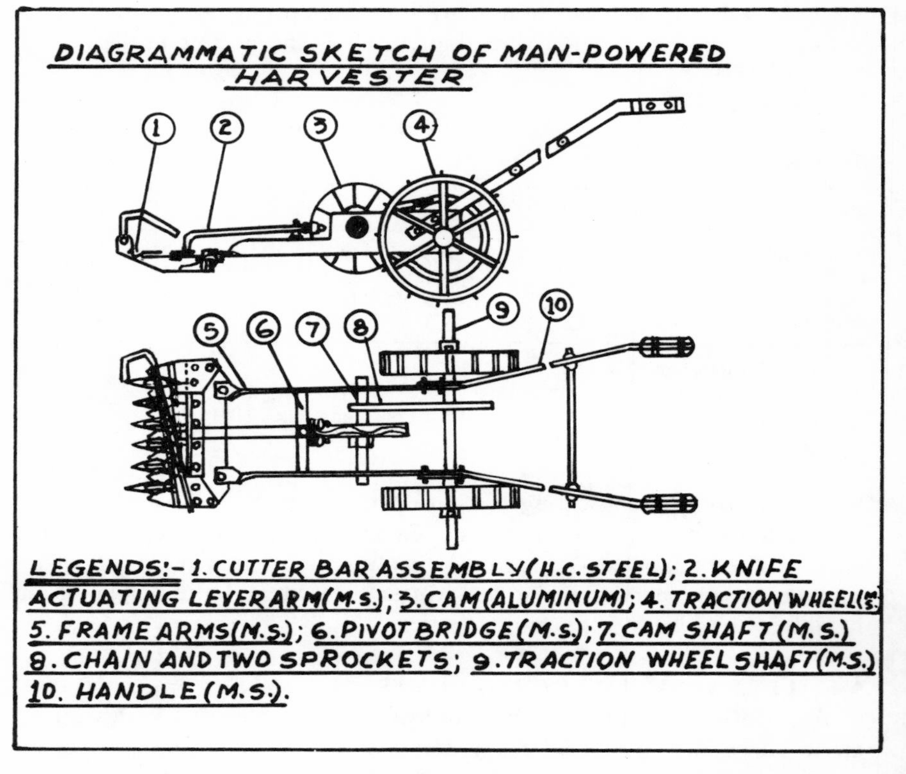

Fig. 11. Diagram of the harvester redesigned by Saran and Ojha.

commodation in this country. While India has abundant time and labor, capital to buy modern technology is seriously lacking. These are the basic conditions that apply to the construction and utilization of Indian buildings (Daftuar 1971*a*). Sinha's design meets at least a part of the problem.

SUMMARY

This paper consists of three principal sections: (1) A discussion of the role of human engineers in facilitating social change in developing countries, with special reference to India; (2) three experimental reports which deal specifically with the theme of the symposium; and (3) a brief survey of Indian work in the area of human factors engineering, with special reference to designs that suit the Indians. Developing countries offer a new field for human factors activities because the effective reapplication of existing human factors knowledge to different levels of technology in newer physical and human environments may require that existing principles be retested and perhaps modified. In so doing, human factors engineers may be better able to bring about desirable social changes throughout the world.

ACKNOWLEDGMENTS

This work has been supported by the Council of Behavioral Research, Maharani Road, Gaya, India. Thankful acknowledgments are due to the Magadh University, Bodh-Gaya, to the Scientific Advisory Council of the North Atlantic Treaty Organization, and to the Indian Council of Social Science Research, New Delhi, for financial support to attend the symposium in Oosterbeek. I am also indebted to J. K. Sinha for his help in data collection, and to S. S. Jha, C. Saran, T. P. Ojha, and N. C. P. Sinha, to Mrs. J. K. Dhesi, and to the Human Factors Society for their help in various ways and their kind permission to reproduce some of the figures and diagrams used.

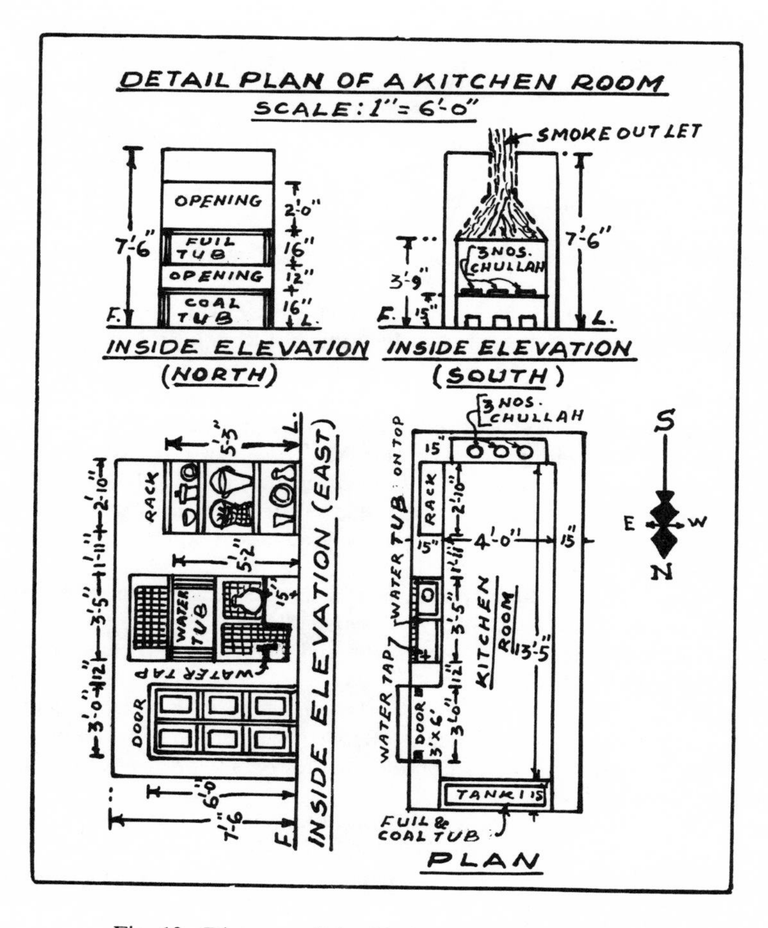

Fig. 12. Diagram of the kitchen designed by Sinha.

REFERENCES

Alexander, R. J. *A primer of economic development.* New York: Macmillan, 1962.

Aziz, A. The legibility of isolated letters of Bengali alphabets. *Indian Psychological Review*, 1970, **6**, 96–99.

Baumgartel, H. Special review of Indian personnel psychology. *Personnel Psychology*, 1966, **19**, 85–92.

Brown, I. D. Human factors in the control of road vehicles. *Electronics and Power*, 1968, **14**, 275–79.

Chakravarty, M. K., and Guharay, A. R. Aerobic work capacity of Indian miners. *Proceedings of the International Biological Symposium*, New Delhi, 1965.

Chatterjee, A., and Daftuar, C. N. Application of Corbusier's human scale to the layout of workspace for typewriting. *Journal of Engineering Psychology*, 1966, **5**, 54–62.

Daftuar, C. N. An application of Modular in a specific human engineering problem. Postgraduate diploma thesis in Industrial Psychology and Industrial Relations (D.I.I.T.), Indian Institute of Technology, Kharagpur, India, 1964.

Daftuar, C. N. Status of industrial psychology in India: a review of published literature. *Indian Psychological Review*, 1969, **5**, 166–83.

Daftuar, C. N. Some psychophysical problems for building designers: a human engineering point of view. *Indian Journal of Psychology*, 1971, **46**, 163–71. (a)

Daftuar, C. N. Human factors research in India. *Human Factors*, 1971, **13**, 345–53. (b)

Daftuar, C. N. Paralogs of zero association value in Devenagari script. *Psycho-Lingua*, 1972, **2**, 83–87.

Daftuar, C. N., and Bhan, V. M. Postural efficiency with reference to paddy workers. Unpublished report, Indian Institute of Technology, Kharagpur, India, 1966.

Datta, A. India. In Pepelasis, A., Mears, L., and Adelman, I. (Eds.), *Economic development: analysis and case studies.* New York: Harper and Row, 1961.

Dhesi, Jagjit K., and Firebaugh, Francille M. The effect of body positions and angles of body bend on heart rate. *Behaviorometric*, 1972, **2**, 1–5. (a)

Dhesi, Jagjit K., and Firebaugh, Francille M. The effect of body positions on angles of body bend during chapati making at the ground-level kitchen. *Indian Journal of Home Science*, 1972, **6**, 10–16. (b)

Fraser, T. M., Jr. Sociocultural parameters in directed change. In A. H. Neihoft (Ed.), *A case book of social change.* Chicago: Aldine, 1966.

Hanson, J. A., and Jones, F. P. Heart rate and small postural changes in man. *Ergonomics*, 1970, **13**, 483–87.

Hansson, J. E., Lindholm, A., and Birath, H. Men and tools in Indian logging operations: A pilot study in ergonomics. Report Number 29. Stockholm: Institutionen för Skogsteknik, Royal College of Forestry, Department of Operational Efficiency, 1966.

Hodge, D. C. Legibility of a uniform-strokewidth alphabet: I. Relative legibility of upper and lower case letters. *Journal of Engineering Psychology*, 1962, **1**, 34–46.

Hodge, D. C. Legibility of a uniform-strokewidth alphabet: II. Some factors affecting the legibility of words. *Journal of Engineering Psychology*, 1963, **2**, 55–67.

Howell, W. C., and Kraft, C. L. Size, blur, and contrast as variables affecting the legibility of alpha-numeric symbols on radar type displays. USAF: WADC TR 59-536, 1959.

Mackie, A. M. A national survey of knowledge of the new traffic signs. Report Number 51. England, Ministry of Transport, Road Research Laboratory, 1966.

McClelland, D. C. *The achieving society*. Princeton, N.J.: Van Nostrand, 1961.

McClelland, D. C., and Winter, D. G. *Motivating economic achievement*. New York: Free Press; 1969.

Pepelasis, A., Mears, L., and Adelman, I. (Eds.). *Economic development: analysis and case studies*. New York: Harper and Row, 1961.

Public Vehicle Department. *Study guides for motor drivers and learners*. Calcutta, 1962.

Saran, C., and Ojha, T. P. Hand operated grain harvester—an aid to small scale mechanization. *Agricultural Engineering*, 1967, **48**, 502–3.

Sen Gupta, A. K., and Sen, R. N. Body measurements of male workers in textile mills. (Tech. Rep. No. 1.) Bombay, India: Central Labour Institute, 1964.

Sinha, D. Psychological researches. In Singh, V. B. (Ed.), *Labour research 1965*. Bombay: Popular Prakashan, 1970.

Sinha, D. Industrial psychology: a trend report. In Indian Council of Social Science Research (ICSSR): *Survey of research in India*. Bombay: Popular Prakashan, 1972.

Tichauer, E. R. The training of industrial engineers for developing countries. *Journal of Industrial Engineering*, 1963, **14**(3), ix–xiii.

SEVEN

Ergonomic Problems in the Transition from Peasant to Industrial Life in South Africa

C. H. WYNDHAM

Since 1950 there has been a marked change in the South African economy from an agrarian to an industrial basis. This change has been accompanied by large numbers of indigenous Bantu moving from a rural, subsistence existence to an urban, industrial life and creating many problems, some of an ergonomic nature common to those in other developing countries. These problems can be grouped under the following headings:

- Anthropometric problems: design of equipment, lay-out of work spaces, design of protective clothing.
- Medicophysiological problems: capacity for physical work as affected by nutrition and endemic diseases in the population.
- Psychological problems: aptitudes, abilities, and skills; training; visual acuity and perception of information; design and layout of controls; decision-making and memory storage.
- Sociocultural problems: cultural limitations; work attitudes and motivations; individual, family and group aspirations; language and communications.

Fairly extensive research into most of the above ergonomic problems has been carried out by the Human Sciences Laboratory of the Chamber of Mines of South Africa and by the National Institute for Personnel Research of the Council for Scientific and Industrial Research in Johannesburg. Some of these subjects also formed part of the multidisciplinary

C. H. Wyndham, Director, Human Sciences Laboratory, Chamber of Mines of South Africa, and Honorary Professor of Environmental and Work Physiology, University of the Witwatersrand, Johannesburg, South Africa.

study of the Bantu in transition from a peasant, rural existence to an urban, industrial life. The latter has been the main research of the Human Adaptability Section of the South African International Biological Programme (IBP). Some aspects of these various researches are given in this paper.

ANTHROPOMETRIC PROBLEMS

Anthropometric surveys carried out by the Human Sciences Laboratory (Morrison, Wyndham, Strydom, Bettencourt, & Viljoen 1968; Strydom, Morrison, van Graan, & Viljoen 1968) have shown that the body dimensions of the South African Bantu male are, in general, significantly smaller than those of males of European origin (Tables 1 and 2). It is important that these differences be taken into account in the design of (a) the spaces in which the Bantu work and (b) the equipment and protective clothing they use, in order to ensure that industrial operations are carried out efficiently and safely. That South African manufacturers are unaware of the differences in functional anthropometry between Bantu and Europeans was revealed in a survey of the protective clothing used by Bantu mineworkers in the gold mining industry.

Table 1. Anthropometric measurements of Bantu mine workers (except for weights, which are in kilograms, all measurements are in mm)

	Percentile				
Dimension	5th	50th	95th	Mean	S.D.
Weight	51.1	59.8	71.2	60.5	5.9
Stature	1,595	1,684	1,792	1,687.5	59.9
Shoulder height	1,305	1,388	1,492	1,391.5	57.7
Elbow height	996	1,065	1,145	1,066.6	44.9
Knuckle height	665	717	775	718.6	33.9
Cristale height	946	1,015	1,100	1,018.2	46.9
Eye height	1,477	1,566	1,670	1,569.2	61.1
Suprasternale height	1,300	1,380	1,472	1,383.4	56.1
Knee cap height	470	509	554	511.0	26.0
Lateral malleolus height	52	63	76	63.0	7.1
Crotch height	750	816	890	818.1	42.9
Overhead reach	1,909	2,023	2,166	2,026.9	77.6
Erect sitting height	813	862	908	860.1	29.8
Normal sitting height	790	838	886	838.7	28.5
Erect sitting eye height	700	747	792	747.3	28.8
Sitting shoulder height	524	566	608	566.5	25.6
Elbow rest height	175	213	253	213.4	23.0
Knee height sitting	497	535	581	536.7	25.0
Popliteal height	380	416	458	416.8	24.4
Thigh clearance height	122	137	159	138.1	10.9
Buttock-knee length	545	588	637	589.5	27.6
Buttock-leg length	942	1,010	1,095	1,013.5	46.9
Seat length	436	471	512	471.4	24.6
Toe to back of seat	670	719	780	721.9	32.8
Functional reach	717	769	850	791.5	40.2
Biacromial diameter	346	377	404	377.1	16.9

Table 1. (*Cont.*)

Dimension	Percentile 5th	50th	95th	Mean	S.D.
Shoulder breadth	394	425	458	426.5	18.6
Chest breadth	254	278	340	278.5	15.2
Hip breadth	287	314	343	314.3	16.8
Waist breadth	238	259	280	258.8	12.4
Maximum body breadth	417	467	510	465.5	26.9
Chest depth	207	224	248	225.8	12.6
Waist depth	187	207	234	208.7	14.6
Buttock depth	194	211	233	212.1	12.1
Neck circumference	318	343	365	342.9	14.2
Shoulder circumference	970	1,045	1,120	1,045.3	43.8
Scye circumference	344	376	411	377.1	19.9
Axillary-arm circumference	240	269	300	268.9	19.3
Elbow circumference	264	288	322	289.7	17.1
Sleeve length	780	834	900	834.9	37.7
Sleeve inseam	450	489	533	491.4	26.1
Inter scye	329	367	410	368.5	25.6
Inter scye maximum	428	467	520	469.1	28.1
Lower thigh circumference	320	355	396	356.3	24.1
Ankle circumference	193	214	235	214.2	12.8
Foot circumference	229	250	272	250.4	13.7
Hand length (third finger)	174	189	205	189.8	10.4
Palm length	99	107	117	108.1	6.1
Hand breadth at thumb	93	101	110	102.4	5.2
Hand breadth at metacarpal	76	84	91	85.2	4.3
Hand thickness	27	31	35	30.9	2.5
Foot length	241	262	282	262.4	12.5
Outer grip	93	103	111	102.8	5.4
Inner grip	43	48	55	48.5	3.5
Crawl height	670	722	796	726.6	40.8
Crawl length	1,240	1,346	1,440	1,342.2	64.9
Prone height	306	354	413	355.6	33.7
Prone length	2,142	2,192	2,333	2,194.8	88.8
Interocular distance	31	36	41	36.0	3.2
Interpupillary distance	60	66	72	66.2	3.4

LEG GUARDS

The main function of leg guards is to protect the shin and knee from flying chips of rock, rolling and falling stones, and abrasions when crawling in narrow spaces on hands and knees. The critical dimension that determines whether the leg guard will fit the leg is the distance between the center of the knee and the top of the foot, which for the middle 90 percent of the Bantu mine workers ranges from 40 to 45 cm. The length of the plastic moulded leg guards from four manufacturers ranged from 35 to 37 cm. Leg guards manufactured for Bantu miners are hopelessly inadequate because they do not fit Bantu legs.

GOGGLES

Gauze goggles are provided for miners working underground to protect their eyes from flying chips of rock. The goggles provided by the manu-

facturers were quite unsuitable. The interocular width of the goggles was 38 mm compared with 36 mm for the 50th percentile of the Bantu. Much more serious was the fact that the interpupillary distance of the 50th percentile for the Bantu was 66 mm, whereas the interpupillary distance of the gauze goggles was 95 mm. These goggles so seriously interfered with their vision that the Bantu did not use them unless forced to do so.

GLOVES

Hand injuries are the commonest cause of lost shifts and have a high rate of temporary and permanent morbidity. The gloves provided by manufacturers were found to be adequate in palm-finger length. This dimension varied from 18.5 to 19.5 cm compared with 20.5 cm for the 50th percentile of the Bantu. The main weakness in the design of the gloves was in the circumference of 26 cm compared with 20 cm for the 50th percentile and 22 cm for the 95th percentile of the Bantu. The 6 cm difference between the circumference of the gloves and of the hands of the average Bantu mineworker is too great, especially for those men who use their hands for manipulating levers and similar kinds of devices.

INDUSTRIAL EQUIPMENT

Similar discrepancies have been found between the dimensions of industrial equipment used by the Bantu and their body dimensions, because

Table 2. Comparison of the means of selected anthropometric measurements for three populations

Dimension	American flight personnel	South African military recruits	Bantu mine workers
Weight	74.4	68.8	60.5
Stature	1,755.4	1,756.4	1,687.5
Crotch height	834.0	846.6	818.1
Sitting height	912.8	917.7	860.1
Popliteal height	431.0	431.1	416.8
Buttock-leg length	1,084.5	1,073.6	1,013.5
Functional reach	812.2	810.3	791.5
Shoulder breadth	454.1	453.3	436.5
Hip breadth	334.5	342.3	300.3
Shoulder circumference	1,149.4	1,093.5	1,045.3
Chest circumference	985.5	907.6	878.9
Waist circumference	813.9	752.5	749.3
Buttock circumference	959.7	925.7	860.5
Thigh circumference	586.6	521.7	496.1
Calf circumference	365.7	395.8	340.9
Hand breadth at metacarpal	88.3	87.7	76.0
Foot length	277.7	265.1	262.4
Foot breadth	96.5	101.0	101.4

even today in South Africa most industrial equipment is designed on the body dimensions of males in UK or Europe. Examples of the differences between these two ethnic groups in certain critical anthropometric measurements are given in Table 3. Although the buttock-leg length of the Bantu is 6 cm shorter than that of the European, the seat-pedal distances in most heavy-duty trucks in South Africa are built on European dimensions. The average Bantu driver who has to drive such vehicles in heavy traffic becomes fatigued because his legs are always operating at a mechanical disadvantage. It is also possible that the Bantu is more liable to accidents than is the European because the former cannot exert proper forces on foot pedals and because his foot can more readily slip off the pedals.

MEDICOPHYSIOLOGICAL PROBLEMS

Most industries in developing countries are labor-intensive. Because, in general, they lack the capital for highly automated machinery, factory operatives are called upon to do more physical work than in the industrially more advanced countries. It is therefore of great importance for industrialists in developing countries to have information about the capacities for physical effort that characterize indigenous populations. With such information, work standards can be set in accordance with the physical work capacities of the workmen.

The Human Sciences Laboratory has been heavily committed to this problem because the mining industry, with which the Laboratory is mainly involved, is labor-intensive and draws its labor force from rural regions all over Southern Africa. Among other things we have measured the maximum oxygen intakes of samples of young Bantu male recruits to the gold

Table 3. Some anthropometric dimensions (in cm) of Bantus and Europeans

	Bantu		European	
	50%	95%	50%	95%
Sitting eye-height (eye-height above the seat is important to provide unobstructed view for locomotive and other drivers)	74.7	79.2	79.4	85.8
Elbow height (this is the best single height for controls)	106.5	114.5	109.1	117.1
Buttock-leg length (the distance between back of heel and rear point of buttock is a key dimension for men operating foot pedals in the sitting position)	101.0	109.5	107.1	115.1

mines and of young European recruits to the Army (Wyndham, Strydom, & Leary 1966). The measurements have been made on the treadmill, unless otherwise stated. The following are some representative results:

Ethnic group	N	Age (years)	Body weights (kg)	VO_2 max (liters/min)	VO_2 max (ml/kg/min)
European	80	20	66.4	3.15 ± 0.407	47.7 ± 3.78
Bantu	90	18–24	60.1	2.86 ± 0.417	46.9 ± 3.96

Although the absolute aerobic power of the European is higher than that of the Bantu, the two groups do not differ significantly when the VO_2 max is computed per kg of body weight. The VO_2 max values of the Bantu were validated on another larger sample of 345 recruits who gave a mean of 2.63 ± 0.452 litres/min (Wyndham 1968). No differences were found in mean body weight and VO_2 max of different tribal groups in Southern Africa (Wyndham, Strydom, Morrison, & Heyns 1966).

The diet of the Bantu in the homelands is deficient in animal protein and marginal in calories. When the Bantu male comes to work in the gold mines he is provided with a diet containing over 4,000 kcal per man per day and 65 gm of animal protein per day. Unlike the pattern of work in the homelands, he also carries out daily physical work at a moderate rate of energy expenditure. As a result of the much improved diet and the regular physical work most Bantu miners show a gain in weight and an increase in VO_2 max. In a sample of men the mean gain in weight over a period of one month was from 55.4 to 58.2 kg. During the same period VO_2 max increased from 2.32 to 2.79 litres/min or from 42.1 to 49.0 ml/kg/min (Wyndham, Strydom, Morrison, Maritz, & Ward 1962).

Similar findings were obtained in the Human Adaptability study for the IBP. This study was done on two groups of Bantu males between the ages of twenty and forty years in rural and urban areas of South Africa. One group consisted of Pedi males in Sekukuniland, the other Venda men in Vendaland. Both groups are in the Eastern Transvaal, but the fertility of the soil is different in the two regions. The results on the two groups were so similar that only those of the Pedi are given here:

	Rural	Urban
N	202	223
Mean age (yr)	32	34
Mean weight (kg)	56.2	60.6
VO_2 max (liters/min)	2.104	2.531
VO_2 max (ml/kg/min)	37.6	41.9

It is clear that the males in urban areas are heavier and have higher VO_2

max's than do those in rural areas. Looked at in terms of their capacities for manual work the comparison between rural and urban males is even more striking:

	Rural (%)	Urban (%)
Light work category (less than 30 ml/kg/min)	16.8	9.0
Moderate work category (between 30 and 45 ml/kg/min)	65.4	57.0
Hard work category (above 45 ml/kg/min)	17.8	34.0

The percentage of urban males capable of hard physical work is almost exactly twice that of rural males. The mean VO_2 max of the mine recruits is higher than that of the rural males in the IBP study because mine recruits are subjected to a medical screening and men less than 50 kg in weight are generally not recruited. The rural sample was chosen at random.

The above results need to be borne in mind in considering the employment of the rural Bantu when he first enters the urban area. Because he is unskilled, he has to take a job as a manual laborer. Yet, as this study reveals, only a small percentage of rural males are capable of heavy manual work. If these men are not to be overtaxed, two steps are necessary. First, the work standard expected from these men should be modified. This applies particularly to the wholly unrealistic work-study standards that are based on Europeans who have higher maximum oxygen intakes. Second, ergonomic principles should be used where possible to lessen the work loads on the men.

There are two possible approaches to the problem of work standards. One is to take an average man and to set work standards in relation to his physical work capacity. In this case it would be an oxygen consumption of 1.3 liters/min, that is, half of the maximum oxygen intake which most physiologists would accept as a reasonable level for shift work. The difficulty is that approximately half of the work population would fall below this level if the distribution of maximum oxygen intake in the population is normal. A more preferable approach is to survey the energy costs of the different tasks in the industry and then use a selection procedure to ensure that only those men with adequate maximum oxygen intakes are put onto the harder work tasks. This is the approach which the Human Sciences Laboratory has adopted in the gold mining industry in South Africa. An extensive survey (Morrison, Wyndham, Mienie, & Strydom 1968) of the oxygen intakes of various mining tasks revealed that the tasks of shoveling rock and tramming it have the highest rate of oxygen intake, that is, 1.4

liters/min. The Laboratory has, therefore, introduced a physical work capacity test by means of which only men with maximum oxygen intakes of 2.8 liters/min and above are selected for these two tasks. Approximately 40 percent of new recruits to the gold mines are classified by the work capacity test as adequate for these hard work tasks.

It is also apparent from these results that consideration should be given to improving the physical work capacities of rural males. This could be done by improving (a) their nutrition through the addition of more calories and animal protein and (b) their health through the eradication of such endemic diseases as malaria and bilharzia. It is unlikely that the health and welfare of the rural populations will be improved solely by their own efforts. Mechanization of agriculture and the use of fertilizers and high-yield crops could radically change the situation, but this would require capital and the education of the population.

PSYCHOLOGICAL PROBLEMS

A great deal of pioneering work was done by the National Institute for Personnel Research in the 1950s on the development of aptitude and ability tests for the selection of Bantu males to undergo training in mechanical tasks and leadership. Recently some of the basic premises upon which the tests were based have been reexamined and new, more sophisticated tests have been introduced (Grant 1970). The Institute was also involved in the development of training programs.

Two research areas of ergonomic importance that have been explored in the South African Bantu are vision and perception.

VISION

In planning displays of information, control panels and the like, it is essential to know if the visual acuity of the workmen is normal, how well they dark adapt, and what percentage of them are color blind (Van Graan, Greyson, Viljoen, & Strydom 1971).

Visual acuity. In a sample of 500 recent Bantu recruits to a mine it was found that, at an illumination level of 1,000 lux, 91 percent had 6/6 vision, 6 percent had 6/9, 2 percent had 6/12 vision and only 1 percent had less than 6/12 vision. Six-twelfth vision is regarded as defective. These figures are better than those obtained by Hopkinson (1949) for a random sample of London school children in which only 83 percent had better than 6/9 vision.

There was a marked reduction in visual acuity with decrease in illumination for both light- and dark-adapted subjects. The decrease was more marked for subjects with less than 6/6 vision than for those with 6/6 vision. Taking the sample as a whole, the percentage of dark- and light-adapted persons with 6/6 vision fell from 91 percent at 1,000 lux to 61 percent at 22 lux and to 29 percent at 5 lux. Twenty-nine percent of the subjects were an exception to this rule and showed no decrease in visual acuity with decrease in illumination. It was also observed that between illumination levels of 1,000 and 22 lux, the visual acuity of light-adapted subjects was better than that of dark-adapted persons, but below 22 lux the reverse was true.

Dark adaptation. In a study of sixty Bantu male recruits to the gold mines each man's eyes were first fully light-adapted by reflected white light. As a further refinement only those with 6/9 or better vision were tested. The population figures were derived from a curve fitted to the individual data following the procedure suggested by Hecht (Jayle, Ourgaud, Baisinger, & Holmes 1959). This gave a mean dark-adaptation time of 33 min, with a range of from 11 to 105 min. The majority of the sample was dark-adapted between 26 and 31 min, but 21 percent took longer than 40 min to dark-adapt fully. According to Brigadier H. C. Nieuwoudt, director of the Institute of Military Medicine in Pretoria, military personnel of the same age range as our Bantu take 29 min to dark-adapt on the average (with a range of 18 to 26 min) after being light-adapted at 1,000 lux.

The Bantu subjects were tested immediately after arrival from their homelands and it was thought that the significantly slower dark adaptation of these men might have been related to the fact that there had been a drought in the homelands for some years resulting in diets deficient in vitamin A. To test this hypothesis, serum vitamin A and carotene levels were measured on seventy-eight recruits to the gold mines immediately on arrival and again three months after engagement. The initial mean vitamin A level was 28.1 mgm percent, with 26 percent of the sample below the minimum acceptable level of 20 mgm percent. Mean carotene level was 93.8 mgm, with 9 percent of the sample below the minimum acceptable level of 40 mgm percent. After three months on the excellent mine hostel diet, the mean vitamin A level was 20.0 mgm percent and 40 percent of the sample was below the minimum acceptable level; the mean carotene level was 45.4 mgm percent and 37 percent of the sample was below the minimum acceptable level. Twenty-four percent of the sample dark-adapted more slowly three months after engagement than on the initial test at recruitment. These results suggest that either the Bantu diet, even in the excellent mine hostel, is deficient in vitamin A or that a high proportion of these men suffer from some subclinical form of liver disease, possibly associated with the high incidence of nutritional diseases in early childhood. Cirrhosis of the liver is much more common among Bantu than among Europeans in South Africa.

Color vision. For European populations the commonly accepted figure for the incidence of color blindness is 8 percent. Only 7 of the 500 Bantu mine recruits (1.4 percent) were found to be colorblind. This much lower incidence of color blindness of the Bantu males was validated in the study of the Human Adaptability Section of the IBP of 350 Venda males where the incidence was found to be 1.0 percent (personal communication from

Professor Herta De Villiers, Department of Anatomy, University of Witwatersrand).

Perception. It is only in recent years that the influence of culture upon pictorial perception has been recognized. This has important ergonomic implications in terms of the display of information such as safety posters and control data on panels. If there are fundamental differences among populations in different phases of acculturation, these need to be taken into account in the presentation of information.

(1) Depth perception, relative size, overlap, and perspective. In 1960 Hudson completed the first of a series of studies on pictorial perception by Africans. The objective was to examine the responses of various cultural groups in Africa to the representational cues of depth in composite pictures. Outline drawings and a photograph of a model scene were constructed so as to depend perceptually on the cues of object size, overlap, and perspective. The three objects in the pictures, hunter, antelope, and elephant, were drawn in a constant relationship to one another in all outline drawings.[1] The first two objects, hunter and antelope, were drawn so as to be perceived as lying in the same frontal plane with the elephant in the background. Supplementary depth cues were introduced in the form of a hill on which the elephant was standing, contour lines representing mountains, and perspective lines representing a road vanishing in a horizon. In all pictures the hunter's spear was aligned with both elephant and antelope. Responses to the questions: (1) Is the hunter aiming at the elephant or the antelope? (2) Is the elephant or antelope nearer the hunter? were taken as self-evident indications of two-dimensional or three-dimensional pictorial perception. There were ten samples in the original study. A white primary school group formed three samples, viz., beginners, finishers, and the total group. A black secondary school group contained two samples, viz., entrants and senior pupils. One group of black graduate school teachers was tested. The remaining four samples were all adult, consisting of (1) a white laborer sample, educated to primary school level, (2) a black illiterate laborer sample, (3) a black laborer sample at primary school level and (4) a black clerical group with secondary school education.

Results showed that white pupils at the beginning of primary school had difficulty in seeing pictures in depth. Even with a photograph, more than one quarter of them perceived in two dimensions. By the end of the primary school the majority of the white pupils were proficient in three-dimensional perception of pictorial material. They had no difficulty at all in seeing the photograph in depth. The first conclusion followed that during the primary school period, the white pupil acquired competence in pictorial depth perception. But the same conclusion did not appear to hold

[1]*Editor's note*: See Figure 5 (page 174) in the article by Sinaiko.

for black pupils. Although the black senior secondary school pupils demonstrated slightly better three-dimensional performance than did black pupils beginning secondary school, the graduate teachers showed no improvement over the senior pupils. In fact, statistical analysis showed that there was no real difference among the three samples of black pupils. That was an important enough finding in itself, but what was still more important was that the depth-perception performance of the more highly educated black samples was not significantly better than that of the white school children in the upper classes of primary school. In view of the difference in educational level, a perceptual difference in favor of the black samples might have been anticipated. In this case, also, all the black graduates had attended multiracial universities and had taken courses and examinations common for both black and white students. These findings compelled the conclusion that formal education, which had been hypothesized as playing a decisive role in the growth of pictorial depth perception, had no more than a contributive function and was subordinate to other cultural factors in the environment.

Further analysis of the occupational samples led to another adjustment in hypothesis (Hudson 1962). Both illiterate and primary educated black laborers saw the pictures flat. So did the white laborers who had primary schooling. It is not difficult to understand that the illiterate black sample should have failed to see depth in pictures. But in the case of the white sample the performance cannot be explained away on ethnic grounds. Neither can it be accounted for educationally, for the white laborer sample had completed primary school. In this instance the factor of cultural isolation tended to nullify the effect of schooling. On the one hand there was a group from an alien culture, on the other hand there was a sample that had virtually rejected its own culture or belonged at best only marginally. Perceptually, the result was the same. Pictures were seen two-dimensionally. Hence formal schooling in the normal course is not the principal determinant in pictorial perception. Informal instruction in the home and habitual exposure to pictures play a much larger role.

On four of the test pictures, evidence on additional ethnic samples was obtained (Hudson 1968). In addition to the European (white) and Bantu (black) samples previously studied, one sample each of Colored (mixed blood) and Indian (Asiatic) pupils with comparable educational standards was tested. A much higher percentage of European pupils saw three-dimensionally. Colored pupils were superior to Bantu and Indian samples. Bantu and Indian pupils performed equally poorly. For consistency of response over three outline pictures the order was European, Colored, Bantu, and Indian samples with more than two-thirds of the European and fewer than one-third of the Indian pupils perceiving three-dimensionally.

Results on the test photograph corroborated this finding. They showed that European primary school pupils were markedly superior to all other cultural groups in the three-dimensional perception of a photograph. This conclusion supports the earlier contention that the perceptual determinants are cultural. The Colored sample is culturally closer to the Western norm than the Bantu sample. It is surprising, however, that the Indian sample had such difficulty and performed at the same level as the Bantu sample. It is worth noting here that the Indian sample, mainly Gujerati-speaking, belong to an alien culture, which, unlike that of the Bantu, has its own distinct form of pictorial art with a set of representational conventions peculiar to the Orient.

Although the prime data in these studies could be quantified and analyzed statistically, responses concerned with the identification of items in the pictures showed additional evidence of misperception of the conventional pictorial depth cues. The outline of the hill on which the elephant stood was seen by the two-dimensional perceivers as a path or a river. Perspective lines were seen to be an elephant trap or poles. The more educated sometimes saw them as letters of the alphabet. Many of the illiterate Bantu sighted the pictures from side to side to determine accurately the direction in which the hunter's assegai was aimed. A limited number rejected their two-dimensional percept on logical grounds, claiming that a hunter would never attack an elephant with a spear only. They concluded that the hunter's assegai must therefore be aimed at the antelope. Bantu graduates reported that they could see the picture in two ways, that is, either flat or in depth, and asked the tester for guidance in deciding their responses.

(2) Foreshortening. Foreshortening is a common convention in Western art and represents a form of perspective applied to the figures of humans and animals generally. Three outline drawings were used in the study. One showed the back view of a man stepping up on a step with one leg. The second presented a back view of a man showing upper arms as far as the elbows. The lower part of his body was cut off by the margin of the picture. In the third plate two plan views of an elephant were drawn. In the one case, the legs and trunk were not visible. It represented a true photographic or foreshortened view of an elephant seen from above. In the other case, all four legs were shown spread out as if the elephant had been flattened out from above. This representation was entirely unnatural and was in fact perceptually impossible.

The three pictures were shown to two samples, one of white primary school pupils, the other of illiterate black laborers. Differences in perception were marked. Most of the white pupils saw the two foreshortened drawings of the man in depth. They reported him to be climbing up a step or to be doing something with his hands, which they could not see because his back hid this from their view. The illiterate black sample saw these two pictures flat. In both cases the figure of the man was seen as maimed or injured in some way. In the first case the man stepping up was seen as having a short leg or a broken leg. In the second picture they reported him with a deformed arm, and sometimes as dead, because he had been cut in half (only the top half of his body having been drawn in).

To the drawings of the elephants the white children responded as we had expected. The foreshortened view of the elephant from above was accepted as natural, while the other view showing all four limbs extended was seen as the skin of a dead elephant. The black illiterate sample reported the precise opposite. They saw the foreshortened elephant as dead, since it had no legs. The second drawing (spread-eagled elephant) was seen as that of a live elephant, in spite of the fact that it was perceptually impossible. One of the illiterate candidates tested reported this to be a very ferocious and dangerous elephant indeed, since it was jumping wildly about.

(3) Twisted perspective. In the representation and identification of perceptually impossible objects there appear to be certain philogenetic and ontogenetic parallels. In his cave drawings of fauna, early man used a form of perspective which has become known as Lascaux or twisted perspective. This is a combination of profile and frontal views. Although the body of the antelope is drawn in profile, two horns and two ears are clearly visible. Drawings of cows collected from black students attending an art school in Rhodesia present the same phenomena. They were asked to draw a cow in profile. Every one drawn had two horns, two ears and four cloven hooves visible. A number of drawings done for a poster competition by black secondary school pupils provided examples of the extension of twisted perspective to inanimate objects such as automobiles. Motor cars drawn in side view showed also the front grille with two head lamps visible. In his early attempts at depicting the human figure the young white child draws full-face views that gradually change to profiles as he grows older. But there is an intermediate stage where characteristics of the early stage occur in the latter. For example, the profiles are drawn with a pair of eyes visible. Pencil drawings of profile views of elephants, collected from illiterate black laborers, show similar phenomena. Irrespective of the shape or size of the drawings, two eyes and four legs were always shown.

The significance of these drawings lies in their demonstration of two distinct representational practices. The white man draws what he sees. If he sees an animal in profile, his drawing of it is perceptually correct. He perceives that a road appears to narrow as it approaches the horizon. So he draws it in this perspective although he knows his picture is conceptually inaccurate. When he does draw, the unacculturated black man draws to identify an object. If he depicts an ungulate in profile, he knows that the creature has cloven hooves, and he includes this item in a perceptually impossible composition. He draws what he knows and not what he sees.

(4) Absolute object size. These drawings of the illiterate black sample gave further interesting pointers to the problem of pictorial perception. One figure only (elephant profile) was drawn on a quarto-sized page. The

range of size was striking. The white man often uses a difference in size in a drawing to indicate that the smaller object is more distant than the larger one. Moreover, the difference in size is directly related to the distance between the two objects. It is clear from the drawings of the black man that he does not make use of this Western perceptual convention to indicate depth and the distance between objects in a figure.

(5) Positioning and orientation. In Western art, objects tend to be positioned with reference to a base line at the bottom of the page. The drawings of the black illiterates were scattered randomly over the page, and it was clear that the base line held no special significance for them as a reference point. Nor did the orientation of their drawings obey Western art conventions. Normally, objects are drawn standing upon a base line toward the bottom of a page, unless the design is inverted. The black illiterates drew elephants rampant, couchant, and even inverted. In the case of inversions, questioning elicited that when they were children these candidates had seen objects and animals being drawn in sand on the ground by an older man. Standing in front of the sand artist, they had always seen things drawn from that viewpoint, with the result that when they came to draw themselves they drew objects in an inverted position quite naturally.

It is clear from the evidence that the perceptual cues and representational conventions common in Western pictorial art provide problems of interpretation to the black man. Differences in objective and lack of meaningful exposure to Western forms of art appear to be largely responsible for these difficulties.

(6) Industrial applications. The problems that arise in the perception of pictorial material by illiterate Bantu workers in industry is illustrated by studies of safety posters carried out on factory workers (Winter 1963) and on mine workers (Hudson, unpublished report).

In the first study 270 black factory workers in South Africa were questioned on their perception and understanding of 6 selected posters. This sample contained workers from urban and rural areas, varying in educational level from complete illiteracy to senior secondary school. The group was subdivided into those who had never before seen the posters, those with some previous vicarious exposure to them, and those to whom the posters had been previously explained. The posters were all printed in color and contained a variety of representational conventions and techniques, for example, single scenes, multiple scenes, serial scenes, cause and effect, and symbolism. Captions printed on each poster to draw attention to the moral of the illustration were covered during initial questioning, but later exposed.

The first two posters were causal scenes of the "before and after" type. The first one was concerned with the wearing of goggles when grinding. It contained four subscenes. The two top scenes showed a worker at the grindstone wearing his goggles and the same worker off duty well dressed and obviously prosperous. The two bottom scenes showed a worker sustaining an eye injury at the grindstone for lack of protective goggles and the same worker now blind and in rags being led around as a beggar. The second poster was concerned with illustrating the danger of stumbling and falling when carrying a load which obscures the view of the worker. Again four scenes were presented, but in this instance the associated subscenes were arranged vertically and not horizontally as in the first poster.

In both cases the black factory workers had difficulty in associating the man in the "before" scene with the same man in the "after" scene. Situation and dress were different and there was no single identifying cue visible. The vertical and horizontal association merely confounded an already complicated perceptual task. In the second poster also there was an interesting cultural problem which the artist had accidentally introduced. The last subscene in the "correct" sequence showed the black worker receiving his wages from the paymaster. In the poster the worker was drawn as receiving his pay with one hand only. For the black factory worker in South Africa this has the opposite connotation. It is customary to receive things in both hands and to give things with one hand. So the moral of the illustration was obscured by an accidental breach of custom; in the poster the recipient of the wage was seen not as a receiver of reward for correct work, but as a donor of money. This was far from the intention of the artist and safety officer.

The third poster showed the danger of throwing tools, for example, a hammer striking another worker on the head. Due to two-dimensional perception by the black worker, the hammer, intended to be seen as in mid-air in the act of striking another worker, was reported as lying on the window sill. With this prime cue missed, there was a search for the message of the poster, and secondary cues tended to be exploited inadvertently. In this poster, for example, all the human figures were shown looking fixedly into space. Hence for some black perceivers the message of this poster was that workers should look at what they are working at and that their attention must not be engaged elsewhere.

Two other posters illustrated the dangers of symbolic convention in Western art. One showed how careless behavior by a worker carrying a heavy wooden plank could injure another worker. The intention of the artist was to portray the actual moment of the accident, for he showed the timber striking the victim on the head and obscuring it entirely. The victim is drawn some distance in the background on a smaller scale than the man in the foreground. The impact is symbolized in the form of a star which surrounds the point of contact of timber and head. The victim, being drawn smaller, is seen as a little boy. The impact mark (the star) is taken to be the sun. In another poster, intended to illustrate the danger of injured toes and fingers as a result of incorrect lifting, impact marks in the form of shock or pain waves were misinterpreted as wrinkles in the man's shirt.

Still another poster was composed to illustrate the danger of standing beneath a loaded crane sling. The artist had drawn a box in a sling of four ropes with one rope broken. Beneath was the figure of the worker with raised arms and petrified with horror, presumably as the box descended upon him. The lower half of the worker in the poster was not shown. The artist had surrounded the poster with a broad red oval border and had highlighted the man's face with reddish brown. This was unfortunate. In the first place, the box was not seen as falling. It was clear that three of the four ropes still held it in place. So the danger has to be seen elsewhere. The color red has acquired a symbolic value for the black worker in South Africa. In large quantities it means fire. In smaller amounts it signifies blood. In the case of this poster it was reported as fire. The man was seen as being in a fiery holocaust. This interpretation was heightened by the highlighting of his features. The deciding factor was the fact that the lower half of his body was not visible. It had obviously been consumed in the flames. So the intended message of the posters was lost due to misperceptions and misunderstandings of custom and pictorial convention.

A number of general conclusions derived from this poster study. Oral explanation improved understanding. Captions made no difference since the bulk of the labor force tested could not read well. Visual exposure to the posters with oral instruction increased understanding. Educated urban workers gave more correct responses than illiterate rural workers.

The findings also produced proposals for the overhaul of conventional poster design. Single scenes were recommended in preference to multiscene posters. Conventional depth cues caused confusion. Symbolic convention caused misunderstanding. The color red had well-established significance and diverted attention from the poster's objective if used for mere decorative purposes. Behavior in the posters should be depicted in accordance with local African custom.

As a result of this study, posters have now been redesigned. Although they may be less attractive from an artistic point of view, they are more effective in promoting understanding. The revised versions of the posters were extensively tested before the designs were finalized and printed. Single scenes only were used. Thus the first multiscene poster was simplified to show one man operating a grinding machine and wearing goggles. There were no "before and after" scenes and no "right-wrong" techniques shown. Great care was taken to show that the worker possessed all his limbs and fingers, even if this meant the loss of perspective and foreshortening. In the case of the "crane sling" poster the color red was avoided entirely. The rope sling itself was shown completely severed, the box which it held was in process of descending and the man underneath was running for his life. In all posters, background and unnecessary details were eliminated. This produced a rather stark representation, which on testing was readily understood. Finally, great care was exercised in the composition of the human figure in all the posters. Throughout, the figure could be identified as the same person. He wore the same clothes and had a constant physiognomy and shade of complexion. This last point was found to be important. When the figure was depicted in a very dark facial tone, the workers rejected him as a foreign worker from Central Africa. He was too dark and they could not identify themselves with him. On the other hand, if his complexion was shown too light, he was seen as a white man. Consequently the message of the poster was not their concern.

In another survey of perception of safety posters, this time in the mining industry, ten posters were shown to three samples. Samples consisted of 81 supervisors, 111 experienced workers, and 100 new recruits. Three of the posters contained two scenes in color and were read vertically, the top scene showing the scene before the accident and the bottom scene depicting the accident or its result. One poster in silhouette contained four scenes read horizontally and was purely instructional in the use of goggles to prevent eye injuries. Analogous symbolism was used in two posters. One consisted of a single scene showing a snake representing a misfire. The other was a dual-scene poster, one scene showing the danger of bad roof conditions underground, the second scene showing an analogous situation from wild life of a bird in front of a trap. Two posters consisted of single scenes. Both were full of irrelevant detail and one relied on foreshortening. One poster was of the minimum presentation type. It contained a colored representation of a pair of hands held up palms facing. One hand had three fingers or parts of fingers missing with one word in the vernacular meaning "no spares." To emphasize the injured hand the artist had drawn blue bands radiating from each amputated finger. The tenth poster included in the test series did not strictly concern itself with safety or accident-prevention. It was intended to illustrate cooperation and consisted of nine scenes reading from top to bottom, depicting the old fable of the dilemma of two calves tied together and faced with the problem of drinking milk from two buckets set farther apart than the length of the rope which bound them.

It was not surprising to find that the samples tested, typical for the industry insofar as they were mostly illiterates, had perceptual difficulties with these posters. Associations with the dual stage posters were not readily understood, so that the message of the poster was distorted or lost. Symbolism frequently proved to be an impediment to understanding, insofar as it distracted attention and distorted the objective of the poster. Since much of the symbolism (for example, snake, bird) was naturalistic, the relevance for mining tended to be lost. The animal represented was misidentified by some of the samples. The snake, a cobra with hood extended, was mistaken for a fish, and the calves were seen as hyenas, dogs, and even lions. More specific conventions intended to illustrate impact, such as shock waves, represented by radiating lines usually accompanied by stars of various colors, were misinterpreted and tended to be taken literally. Where the poster showed an eye accident at the moment of happening, the white stars were taken to be the white parts of the eye, the black stars the black parts (that is, the pupil of the eye) and the red stars blood, according to the common convention. Foreshortening of the human figure produced the response of injury which implies two-dimensional perception. Plethora of detail and economy of presentation both defeated the objective. In the one instance the message was lost in a mass of irrelevant background items. In the other case (the poster showing the amputated fingers) there appeared to be too little information to permit the objective to be grasped. The blue lines with which the artist surrounded the fingers in an attempt at graphic emphasis proved to be confusing and distracting. They were seen as perspiration, muscles, fingernails, and rays of light.

The purpose of each poster was to communicate a specific message appropriate to a specific accident situation. Most of the workers knew that the posters were associated with danger or accidents, but in a vague, general relationship which nullified their specificity. Colored geometric patterns would have made just as suitable posters and would have been nearly as effective as communicators. As might be expected, supervisors showed more understanding of the posters than the other two samples. Raw recruits had greatest perceptual difficulties. Some examples will serve to show how far off the mark understanding was. The poster showing the hands with injured fingers was intended to show that fingers are irreplaceable and that care should be taken to protect hands by working carefully. Misinterpretations of this message were as follows: "The poster teaches me how to write," "The man is counting with his fingers," "It tells of a man who has open arms," "I should go and do likewise" (sic), "I must wash my hands," "I must give hand signals," "I must not handle fire." The poster showing the two calves designed to illustrate the advantages of cooperation gave the following misunderstandings: "It teaches me how to attend cattle," "It shows me how to feed calves," "It shows me how to bring up calves without a mother," "It tells me to rest and drink when tired," "It tells me to work like cattle." So much for naturalistic analogies among a cattle-owning population!

SOCIOCULTURAL PROBLEMS

The degree of acculturation of the rural Bantu is of paramount importance in his ability to adapt successfully to innovation in industry, such as the introduction of ergonomics into the work situation. A sociocultural study (Glass 1964; Wyndham 1965) of 155 Bantu mine recruits from Sotho homelands in South Africa showed that these men were typical of peasants elsewhere in the world in their close adherence to traditional ways of life, such as the lobola, or bride-price, payment on marriage. They are, however, experiencing forces that are leading to cultural change. Although only eight percent were literate (using Standard III as a criterion of literacy) and only 10 percent spoke a European language, 50 percent were members of an orthodox Christian church.

These new cultural forces brought to the surface two recognizably different groups in this population. One is "conservative" and tribally oriented. It comprised approximately 60 percent of this sample. These men have limited individual, family and group aspirations. They only remain in the urban areas sufficiently long to satisfy their need for money to purchase food, clothing, and livestock. They reject outright any suggestion that their families should migrate permanently to urban areas. Their leisure activities, both at home and in the urban areas, are traditional. These men, in consequence, are not interested in learning new skills or in seeking promotion; they cling to the jobs which their tribe has by tradition always had. They prefer to work with associates from their homelands and under a supervisor from the same language group. They are slow to learn a new language and are poor in communication. They display considerable resistance to innovations of any sort. These negative attitudes are an obstacle to the introduction of even simple ergonomic principles in the work situation and have been the reason for the failure, for example, of these men to accept training in better, more ergonomic, methods of shoveling rock in mines. The new methods would have led to a reduction in energy expenditure, to the greater production of rock from stopes, and to higher bonuses!

The other 40 percent of the sample was "forward looking." Men in this category prefer to work in secondary industry in South Africa because of the greater opportunities for learning new skills, for getting promoted, and for earning higher pay. They have generally extended their wants beyond the primitive subsistence level and readily move to urban areas with their families. They have higher aspirations for themselves and for their families. They often are prepared to make sacrifices for the education of their children and are prepared and eager for change; they accept innovations readily.

The successful introduction of new ergonomic principles on the workshop floor depends upon recognizing the differences between these two groups, winning over the forward-looking men to the new procedures and the introduction of new methods through these men.

REFERENCES

Glass, Yeta. The motivation of contract lashers—a social psychological study. Johannesburg: National Institute of Personnel Research Report No. C/Pers 16, 1964.

Grant, G. V. The development and validation of a classification test battery constructed to replace the general adaptability battery. Johannesburg: National Institute of Personnel Research Report No. C/Pers 181, 1970.

Hopkinson, R. G. Studies of lighting and vision in schools. *Transactions of the Illuminating Engineering Society*, 1949, **14**, 149–65.

Hudson, W. Pictorial depth perception in sub-cultural groups in Africa. *Journal of Social Psychology*, 1960, **52**, 183–208.

Hudson, W. Pictorial perception and educational adaptation in Africa. *Psychologia Africana*, 1962, **9**, 226–39.

Hudson, W. The study of the problem of pictorial perception among unacculturated groups. *International Journal of Psychology*, 1968, **2**, 89–107.

Jayle, G. E., Ourgaud, A. E., Baisinger, L. F., and Holmes, W. J. *Night vision*. Springfield, Illinois: Thomas, 1959.

Morrison, J. F., Wyndham, C. H., Mienie, B., and Strydom, N. B. Energy expenditure in mining tasks and the need for the selection of labourers. *Journal of the South African Institute of Mining and Metallurgy*, 1968, **69**, 185–91.

Morrison, J. F., Wyndham, C. H., Strydom, N. B., Bettencourt, J. F., and Viljoen, J. H. An anthropometrical survey of Bantu mine workers. *Journal of the South African Institute of Mining and Metallurgy*, 1968, **68**, 275–79.

Strydom, N. B., Morrison, J. F., van Graan, C. H., and Viljoen, J. H. Functional anthropometry of white South African males. *South African Medical Journal*, 1968, **42**, 1332–35.

Van Graan, C. H., Greyson, J. S., Viljoen, J. H., and Strydom, N. B. Visual acuity of Bantu mine workers. Johannesburg: Chamber of Mines Research Report No. 16/71, 1971.

Winter, W. The perception of safety posters of Bantu industrial workers. *Psychologia Africana*, 1963, **10**, 127–35.

Wyndham, C. H. An operational study of the influence of human and other factors in industrial productivity. *Transactions of the South African Institute of Mechanical Engineers*, 1965, **10**, 239–51.

Wyndham, C. H. An examination of the methods of physical classification of African labourers for manual work. *South African Medical Journal*, 1968, **40**, 275–78.

Wyndham, C. H., Strydom, N. B., and Leary, W. P. The maximum oxygen intake of young, active Caucasians. *Internationale Zeitschrift für angewandte Physiologie*, 1966, **22**, 296–303.

Wyndham, C. H., Strydom, N. B., Morrison, J. F., and Heyns, A. J. The capacity for endurance effort of Bantu males from different tribes. *South African Journal of Science*, 1966, **62**, 259–63.

Wyndham, C. H., Strydom, N. B., Morrison, J. F., Maritz, J. S., and Ward, J. S. The influence of a stable diet and regular work on body weight and capacity for exercise of African recruits. *Ergonomics*, 1962, **5**, 435–44.

EIGHT

Direction-of-Movement Stereotypes in Different Cultural Groups

PAUL VERHAEGEN, RICHARD BERVOETS, GHISLAIN DEBRABANDERE, FILIP MILLET, GUIDO SANTERMANS, MARCEL STUYCK, DIRK VANDERMOERE, AND GUIDO WILLEMS

In this paper we report the results of some exploratory research on direction-of-movement stereotypes or preferences in a few non-Western cultural groups. The choice of the groups was mostly determined by their accessibility. Algerian and Moroccan workers and Moroccan girls and boys were tested in Belgium by two of us (R. B. and G. W.). These Moslem groups were interesting because the Arabic language is written from right to left, a practice which might be presumed to produce habits that interfere with some common Western movement stereotypes. As it turned out, however, 50 percent of our adult Moslem subjects wrote as much French as Arabic. The other 50 percent were only able to write their names and had learned to do this in Latin characters (that is, left to right). In Central Africa, Negro groups of both sexes and of various educational levels were tested at different places by four of us (F. M., G. S., M. S., and D. V.), young psychologists working temporarily as secondary school teachers. These Negro groups were interesting because they have had various degrees of exposure to Western technological culture.

Paul Verhaegen, Department of Psychology, Katholieke Universiteit te Leuven, Leuven, Belgium.

METHOD

The subjects were told that they were participating in a study on movements. Steps were taken to prevent subjects who had been tested from communicating with subjects still to be tested.

Each subject's handedness was determined by four simple tests. Each subject was asked to (1) unscrew the cap of a little bottle; (2) write his name or, if he could not write, draw a small figure and then erase it; (3) cut out with scissors, or with the point of a knife if unfamiliar with scissors, a circle drawn on a sheet of paper; and, finally, (4) throw a ball or a stone at some object 5 meters away. The material was presented in such a way that it could be gripped equally well with either hand. A subject was classified right-handed if he did all four tasks with his right hand, and left-handed if he performed all four tests with his left hand. Subjects who were not consistent on these tests were classified as uncertain.

Half of the right-handed subjects were tested using only their left hands and half using only their right hands (except for one item in which both hands were to be used). There were so few left-handed subjects and so few subjects in the uncertain group that their data have not been analyzed in this paper.

After having been told with which hand to work and to respond at all times in a natural, spontaneous way, the subject was asked to rotate a real knob mounted on a wooden board in any way he liked. We recorded whether he made a clockwise or a counterclockwise movement. The subject was then asked to rotate simultaneously two knobs, one with each hand. These knobs were mounted on a wooden board about half a meter from each other. We recorded the way each knob was rotated.

Next the subject was presented with twelve drawings, each on a separate card, always in the same order. The drawings represented eight vertical scales, each with an associated knob alternately to the right and to the left, and four horizontal scales, each with a knob underneath the scale. In each case a pointer was to be moved in a given direction, as indicated by an arrow on the drawing. The subject had to show with his right (or left) hand how he preferred to rotate the knob to get the desired movement of the pointer. We hoped that the two preliminary test items using real knobs had made the situation clear to the subjects. They were also told that the task was like tuning a radio receiver. In every case we noted whether a clockwise or a counterclockwise movement was chosen. These twelve test items are represented in Figure 1.

A last test card (item 13 in Fig. 1) was designed to represent a moving dial with rotary control. The window showed the digit 1, and the subject had to demonstrate or to say how he would rotate the knob to get the digit 2 to appear in the window. All testing was done individually.

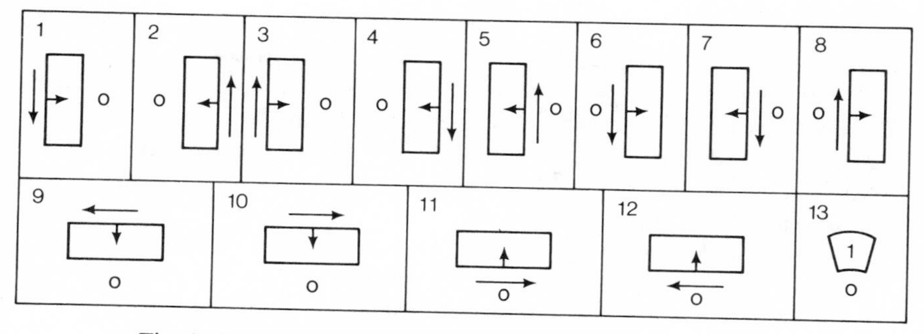

Fig. 1. The 13 test items referred to in Tables 2, 3, 5, and 6.

We recorded each subject's sex, age and educational level. Each subject was also questioned about his familiarity with devices such as radios, bicycles, alarm clocks, watches, sewing machines, and automobiles.

Our tests were very simple. For example, the displays and controls were always presented on the same plane. However, we felt that it was useful to study the reactions to designs for which strong direction-of-movement preferences exist in Western culture. This series of test items had already been used by one of the authors in a study on differences in stereotypes between right- and left-handed workers in an automobile plant (Debrabandere 1971). In that study, stereotypes appeared to be stronger in right-handed than in left-handed subjects. The fact that this simple test material yielded data that confirm those obtained by Chapanis and Gropper (1968) using elaborate instrumentation, argues for the validity of these simple techniques.

SUBJECTS

In describing each group of our subjects we mention the number of not clearly right-handed subjects discovered, that is, left-handed and those with uncertain handedness. These data may, perhaps, have some interest for students of differences in handedness among different ethnic or cultural groups (Verhaegen & Ntumba 1964).

The first group of subjects, in a certain sense a reference group, was composed of 60 right-handed Flemish workers studied in 1969 (Debrabandere 1971). Accurate data on the occurrence of left handedness in Flemish workers are not available from that study.

The other subjects, North Africans and Central African Negroes, were all tested in February and March 1972. Thirty-eight right-handed Algerian subjects out of a total of 41 were tested. The other three were left-handed and were discarded. All these subjects were coal miners who had been in Belgium less than a month at the time of testing. They were all fluent in

French. Their mean age was 27 years with a range of 24 to 35 years. The testing was done on the premises of the mines.

In testing a group of 54 Moroccans the help of an interpreter was essential, even though the subjects had lived in Belgium for a few years. The majority of these subjects were tested at their homes, some on the premises of the coal mine. No left-handed subjects were discovered. The mean age was 32 years with a range of 15 to 44 years.

Twenty-nine Moroccan boys (age range of 6 to 15 years) and 29 girls (age range of 8 to 16 years) were tested at school. They understood Dutch. No left-handed Moroccan children were found.

Some cultural taboos prevented us from testing Moslem women.

In Matadi, port at the Zaïre River (formerly the Congo), we tested 96 right-handed Negro male adolescents or young men, students in the final years of secondary school, with ages ranging from 20 to 22. In addition, 3 subjects with uncertain handedness were found and discarded.

Also in Matadi we tested 87 right-handed Negro men (mean age 37, range of 15 to 53 years). In addition, 2 left-handed subjects and 16 with uncertain handedness were found. Of the latter, 11 came from a group for which one of three African interpreters had been responsible. This suggests that some systematic error may have been introduced by the interpreter. Most of these subjects had not completed primary school. They are best described as semiliterate people.

In Kangu, a little place situated in the Mayumbe forest about 100 kilometers north of Boma, we tested 106 Negro boys, 14 and 15 years old. All were right-handed pupils in their first year of secondary schooling.

In Lokutu (formerly Elisabetha), a small village on the Zaïre River northwest of Kisangani (formerly Stanleyville), we tested 77 right-handed Negro boys, 13 to 18 years of age, pupils in the first years of secondary school. In addition 2 left-handed boys and 8 with uncertain handedness were found and discarded.

In the town of Kamina and the nearby mission of Luabo we tested 60 right-handed Negro men with a mean age of 51 years and a range of 24 to 80 years, and 60 right-handed Negro women with a mean age of 45 years and a range of 21 to 80 years. These men and women were illiterate or semiliterate. Their education had not progressed beyond a few years of primary schooling. There were no cases of left-handedness. In the same places we also tested 101 Negro girls, 15 and 16 years old, and 130 Negro boys, 15 and 16 years old, finishing primary school or starting secondary school. Only one case of uncertain handedness (a girl) was found.

RESULTS

The basic data are given in Tables 1, 2, and 3. For each test item we show the numbers of subjects of each group who turned the knob clockwise (*C*) and counterclockwise (*CC*) or, in the case of the test item requiring the subject to turn two knobs, the numbers of subjects who performed the movements in a parallel, *P*, or in a symmetrical, *S*, way. The significance of the difference between the numbers in each cell was tested with the binomial test (Siegel 1956) using a two-tailed criterion. Note that in these three tables the cell totals in a given column are not necessarily identical. Occasional small discrepancies are due to inadvertent omissions of data by the several experimenters or the failure of a subject to respond to a test item.

Tables 4, 5, and 6 combine the data from the second, third, fourth, and fifth columns of Tables 1, 2, and 3, respectively. This summation was done because the original groups were rather small. In the discussion that follows we shall sometimes use the North African data in Tables 4, 5, and 6, rather than those in the second, third, fourth, and fifth columns of Tables 1, 2, and 3.

KNOB-TURNING TESTS

Table 1 shows that when subjects turned a knob with their preferred right hand, in 12 out of the 13 groups the majority preferred the clockwise direction. This is significant at the 1 percent level for 8 groups and at the 5 percent level for one group. Analysis of the same data, after combining those for the North Africans (Table 4) which reduced the total number of groups to 11, showed that the majority of the subjects in all 11 groups preferred a clockwise movement. This effect is significant at the 1 percent level in 9 of the 11 groups. It is not significant only for the Negro girls and the Negro women.

When subjects turned the knob with their nonpreferred left hands, the results were equivocal. In 7 out of 13 groups, the majority preferred a clockwise direction; in the remaining 6 groups the majority preferred a counterclockwise direction. Even the significant effects were almost evenly divided between the two directions of movement. For 3 groups there was a significant difference in favor of a clockwise movement; in 2 groups a significant difference in favor of a counterclockwise movement. If we eliminate the Flemish, that is, the European group, the results are exactly divided between the 2 directions of movement for the remaining 12 groups of subjects, all Africans. Perhaps the only conclusion that can be drawn from these data is that on this particular task Africans show complete ambivalence about how to respond. Unfortunately for purposes of this comparison, we do not have sufficient data on European groups to enable us to decide whether there is evidence here for a genuine cultural difference.

In the task involving the two knobs it is possible to rotate both knobs clockwise or both counterclockwise, or to turn one knob clockwise and the other counterclockwise. These 4 possibilities have been combined so that

Table 1. Direction of knob-turning movements in different cultural groups

	Flemish workers		Algerian workers		Moroccan workers		Moroccan boys		Moroccan girls		Negro boys: Luabo and Kamina		Negro boys: Kangu		Negro boys: Lokutu		Negro adolescents: Matadi		Negro men: Matadi		Negro men: Kamina		Negro girls: Luabo and Kamina		Negro women: Kamina	
	CC	C	CC	C	CC	C	CC	C	CC	C	CC	C	CC	C	CC	C	CC	C	CC	C	CC	C	CC	C	CC	C
Single knob turned with preferred right hand	*0*	*30*	11	8	*3*	*25*	*1*	*13*	6	9	*18*	*48*	*17*	*35*	*12*	*28*	*10*	*39*	**13**	**28**	*5*	*25*	18	32	13	17
Single knob turned with nonpreferred left hand	**1**	**8**	6	13	12	16	8	7	10	4	31	33	26	28	**25**	**12**	*12*	*35*	*12*	*34*	19	11	29	22	*23*	*7*
	P	S	P	S	P	S	P	S	P	S	P	S	P	S	P	S	P	S	P	S	P	S	P	S	P	S
Two knobs turned with both hands	31	28	**27**	**11**	27	27	8	20	**8**	**21**	**52**	**78**	56	50	*14*	*63*	*62*	*34*	39	48	26	34	50	51	*12*	*48*

Notes: Numbers in boldface within any cell differ at $0.01 < p < 0.05$.
Numbers in italics within any cell differ at $p < 0.01$.

Table 2. Direction-of-movement stereotypes in different cultural groups (subjects working with their preferred right hands)

Test item	Flemish workers		Algerian workers		Moroccan workers		Moroccan boys		Moroccan girls		Negro boys: Luabo and Kamina		Negro boys: Kangu		Negro boys: Lokutu		Negro adolescents: Matadi		Negro men: Matadi		Negro men: Kamina		Negro girls: Luabo and Kamina		Negro women: Kamina	
	CC	C	CC	C	CC	C	CC	C	CC	C	CC	C	CC	C	CC	C	CC	C	CC	C	CC	C	CC	C	CC	C
1	**21**	**9**	14	5	14	11	8	6	6	9	*46*	*20*	32	20	**28**	**12**	*34*	*15*	25	16	19	11	**34**	**16**	14	16
2	19	11	12	7	15	10	7	7	7	8	*46*	*20*	23	29	19	21	*34*	*15*	24	17	18	12	22	28	15	15
3	**9**	**21**	**4**	**15**	**6**	**19**	7	7	6	9	*13*	*53*	21	31	*6*	*34*	*10*	*39*	16	25	**9**	**21**	*9*	*41*	**9**	**21**
4	**8**	**22**	6	13	12	13	8	6	10	5	*20*	*46*	26	26	*11*	*29*	18	31	20	21	11	19	28	22	17	13
5	**9**	**21**	5	14	*4*	*21*	**2**	**12**	5	10	*13*	*53*	*16*	*36*	*10*	*30*	**17**	**32**	**13**	**28**	**9**	**21**	*9*	*41*	*5*	*25*
6	14	16	6	13	*4*	*21*	8	6	11	4	27	39	**17**	**35**	16	24	*11*	*38*	16	25	13	17	**33**	**17**	17	13
7	*24*	*6*	13	6	**19**	**6**	11	3	11	4	41	25	**35**	**17**	*31*	*9*	*36*	*13*	21	20	15	15	*37*	*13*	17	13
8	17	13	13	6	15	10	5	9	4	11	**42**	**24**	30	22	*29*	*11*	*35*	*14*	21	20	15	15	18	32	11	19
9	*27*	*3*	11	8	17	8	11	3	11	4	*47*	*19*	*44*	*8*	*34*	*6*	*43*	*6*	*31*	*10*	20	10	*36*	*14*	18	12
10	*2*	*28*	*3*	*16*	8	17	4	10	7	8	*15*	*51*	*9*	*43*	*3*	*37*	*6*	*43*	14	27	**9**	**21**	20	30	11	19
11	*5*	*25*	5	14	8	17	6	8	**3**	**12**	*11*	*55*	*9*	*43*	*6*	*34*	*8*	*41*	*6*	*35*	13	17	19	31	**8**	**22**
12	*26*	*4*	13	6	**19**	**6**	**12**	**2**	8	7	*51*	*15*	*44*	*8*	*33*	*7*	*39*	*10*	**29**	**12**	13	17	**34**	**16**	14	16
13			8	11	12	13	7	7	8	7	34	32	28	24	25	15	29	20	20	21	12	18	*13*	*37*	13	17

Notes: Numbers in boldface within any cell differ at $0.01 < p < 0.05$.
Numbers in italics within any cell differ at $p < 0.01$.
On test item 13, most Flemish workers turned the knob clockwise ($0.01 < p < 0.05$). The exact numerical entries are, however, no longer available.

Table 3. Direction-of-movement stereotypes in different cultural groups (subjects working with their nonpreferred left hands)

Test item	Flemish workers		Algerian workers		Moroccan workers		Moroccan boys		Moroccan girls		Negro boys: Luabo and Kamina		Negro boys: Kangu		Negro boys: Lokutu		Negro adolescents: Matadi		Negro men: Matadi		Negro men: Kamina		Negro girls: Luabo and Kamina		Negro women: Kamina	
	CC	C	CC	C	CC	C	CC	C	CC	C	CC	C	CC	C	CC	C	CC	C	CC	C	CC	C	CC	C	CC	C
1	18	12	**15**	**4**	**22**	**7**	8	7	9	4	*43*	*21*	*41*	*13*	*27*	*10*	30	17	**30**	**16**	**21**	**9**	*37*	*14*	**23**	**7**
2	20	10	**15**	**4**	**21**	**8**	9	6	6	7	37	27	31	23	21	16	28	19	29	17	18	12	25	26	17	13
3	11	19	*2*	*17*	14	15	7	8	7	6	**23**	**41**	*16*	*38*	**11**	**26**	18	29	20	26	12	18	**17**	**34**	17	13
4	**8**	**22**	*1*	*18*	11	18	7	8	5	8	*19*	*45*	*16*	*38*	15	22	**15**	**32**	23	23	**9**	**21**	**18**	**33**	13	17
5	*4*	*26*	**4**	**15**	**8**	**21**	10	5	7	6	**22**	**42**	**19**	**35**	*9*	*28*	*10*	*37*	19	27	11	19	*15*	*36*	14	16
6	*5*	*25*	*2*	*17*	9	20	6	9	5	8	**22**	**42**	**18**	**36**	*10*	*27*	**16**	**31**	19	27	12	18	**17**	**34**	12	18
7	**21**	**9**	11	8	18	11	**12**	**3**	10	3	*43*	*21*	30	24	*30*	*7*	*36*	*11*	26	20	18	12	30	21	13	17
8	*23*	*7*	13	6	**21**	**8**	**12**	**3**	6	7	40	24	*40*	*14*	**25**	**12**	**31**	**16**	27	19	**21**	**9**	**33**	**18**	19	11
9	*28*	*2*	14	5	*24*	*5*	11	4	10	3	*50*	*14*	*43*	*11*	**26**	**11**	*38*	*9*	**31**	**15**	**22**	**8**	*36*	*15*	18	12
10	*6*	*24*	*2*	*17*	11	18	6	9	4	9	*16*	*48*	*10*	*44*	**7**	**30**	*8*	*39*	**15**	**31**	10	20	*10*	*41*	16	14
11	*4*	*26*	5	14	10	19	6	9	6	7	*14*	*50*	*16*	*38*	*10*	*27*	*3*	*44*	17	29	14	16	*10*	*41*	11	19
12	*27*	*3*	14	5	**22**	**7**	*14*	*1*	10	3	*48*	*16*	*43*	*11*	*30*	*7*	*43*	*4*	*33*	*13*	19	11	*44*	*7*	*25*	*5*
13			9	10	18	11	11	4	7	6	30	34	**35**	**19**	*28*	*9*	28	19	18	28	13	17	26	25	17	13

Notes: Numbers in boldface within any cell differ at $0.01 < p < 0.05$.
Numbers in italics within any cell differ at $p < 0.01$.
On test item 13, most Flemish workers turned the knob clockwise but this preference was not significant at the 5 percent level. The exact numerical entries are no longer available.

in Tables 1 and 4 the only comparison made is that between parallel and symmetrical movements. For this task there is a slight tendency for most groups to favor symmetrical movements (8 out of 13 groups). In 4 groups the majority preferred parallel movements and in one group the data are evenly divided between the two alternatives.

From these data, the following generalizations seem valid:

1. In turning a single knob with their preferred right hands, almost all our groups preferred a clockwise movement. This appears to be a strong movement stereotype that is consistent across all the cultural groups we tested.

2. In turning a single knob with their nonpreferred left hands our African groups showed complete ambivalence about which direction to turn. There may be evidence here for cultural differences, but we did not, unfortunately, test enough European subjects to be able to reach any firm conclusions on this point.

3. In turning two knobs simultaneously with both hands, there was a tendency for most groups of subjects to use symmetrical movements of the two hands. There was so much disagreement among our several groups that we can classify this as only a weak stereotype.

STRAIGHT SCALES WITH ASSOCIATED KNOBS

General tendencies. One way of finding general tendencies is to look at the data for (a) straight scales with associated knobs, that is, test items 1 through 12, (b) cells that show significant differences at $p \leqslant 0.05$, and (c) African groups only, that is, the 12 groups not including the Flemish workers. In Tables 2 and 3, the numbers in 140 cells satisfy all three bases of classification. The movement preference represented by the data in each of these 140 cells was compared with the preference shown by the Euro-

Table 4. This table combines data from the second, third, fourth, and fifth columns of Table 1

Test item	Algerian and Moroccan workers		Moroccan boys and girls	
	CC	C	CC	C
Single knob turned with preferred right hand	*14*	*33*	*7*	*22*
	CC	C	CC	C
Single knob turned with nonpreferred left hand	18	29	18	11
	P	S	P	S
Two knobs turned with both hands	54	38	*16*	*41*

Note: Numbers in italics within any cell differ at $p < 0.01$.

Table 5. This table combines data from the second, third, fourth, and fifth columns of Table 2

	Algerian and Moroccan workers		Moroccan boys and girls			Algerian and Moroccan workers		Moroccan boys and girls	
Test item	CC	C	CC	C	Test item	CC	C	CC	C
1	28	16	14	15	8	28	16	**9**	**20**
2	27	17	14	15	9	28	16	*22*	*7*
3	*10*	*34*	13	16	10	*11*	*33*	11	18
4	18	26	18	11	11	**13**	**31**	**9**	**20**
5	*9*	*35*	*7*	*22*	12	*32*	*12*	**20**	**9**
6	*10*	*34*	19	10	13	20	24	15	14
7	*32*	*12*	*22*	*7*					

Notes: Numbers in italics within any cell differ at $p < 0.01$; those in boldface differ at $0.01 < p < 0.05$.

pean group, that is, the Flemish workers, on corresponding test items. For purposes of these comparisons, we were interested in whether the African group exhibited the same preference as the majority of the Flemish group, irrespective of whether the difference was significant in the Flemish group. In only 1 of the 140 such comparisons was there a difference between the preferences exhibited by the Africans and those by the Flemish workers. That difference was for the Negro girls on test item 6 in Table 2.

In Tables 5 and 6 there were 27 cells that satisfied all three bases of classification above. In only one of the 27 instances was there a discrepancy between what the Africans and Flemish workers did. That discrepancy was for the Moroccan boys and girls on test item 8 in Table 5.

Taking all these comparisons into account it appears that by and large Africans exhibit the same movement preferences as Europeans.

Table 6. This table combines data from the second, third, fourth, and fifth columns of Table 3

	Algerian and Moroccan workers		Moroccan boys and girls			Algerian and Moroccan workers		Moroccan boys and girls	
Test item	CC	C	CC	C	Test item	CC	C	CC	C
1	*37*	*11*	17	11	8	*34*	*14*	18	10
2	*36*	*12*	15	13	9	*38*	*10*	**21**	**7**
3	**16**	**32**	14	14	10	*13*	*35*	10	18
4	*12*	*36*	12	16	11	**15**	**33**	12	16
5	*12*	*36*	17	11	12	*36*	*12*	*24*	*4*
6	*11*	*37*	11	17	13	27	21	18	10
7	29	19	*22*	*6*					

Notes: Numbers in italics within any cell differ at $p < 0.01$; those in boldface differ at $0.01 < p < 0.05$.

Differences between vertical scales with knobs on the right versus those on the left. It is generally accepted that rotary controls should be placed to the right, and not to the left, of vertical scales (see, for example, Chapanis 1965, p. 110). Our data tend to confirm this rule. For example, when subjects were working with their right hands (Tables 2 and 5) and when knobs were on the right (test items 1, 3, 5, and 7), there were 31 cells in which preferences were significant at $p \leqslant 0.05$, and 13 cells in which differences were not significant. By contrast, for test items with knobs on the left (test items 2, 4, 6, and 8) these numbers were exactly reversed: only 13 cells showed significant differences at $p \leqslant 0.05$, and 31 cells showed no significant differences. When these numbers themselves were tested by chi square the differences were significant at the 2 percent level. Although there was a similar trend in the data collected with the nonpreferred left hand, chi square shows that the trends are not significant.

Differences between vertical and horizontal scales. For each group of subjects the twelve test items were ranked in terms of the strength of the preference shown by that group. The means of the ranks were then computed for all groups of subjects in Tables 2 and 3 (Table 7).

Inspection of Table 7 shows that the mean ranks are lower for horizontal scales (items 9, 10, 11, and 12) than for vertical scales (items 1 to 8). The difference is significant at the 5 percent level (Mann Whitney U-test, two-tailed; Siegel 1956). There is also a difference between the mean ranks for the vertical scales with controls on the right (items 1, 3, 5, and 7) and those with controls on the left (items 2, 4, 6, and 8) in favor of the scales with knobs on the right, but this difference is not significant. The difference between the mean ranks of the vertical scales with knobs on the left (items 2, 4, 6, and 8) and the mean ranks of the horizontal displays (items 9, 10, 11, and 12), is again significant at the 5 percent level.

In the preceding computations the data for the Flemish workers were combined with those of the other cultural groups. Deleting the data for the Flemish workers does not change the results appreciably.

Relationship between the strength of direction-of-movement preferences and the use of radios. For each group we counted the number of cells in Tables 2 and 3 (and 5 and 6) that showed a significance difference at the

Table 7. Mean ranks of the strength of direction-of-movement preferences for 12 scales with associated rotary controls (smaller mean ranks indicate stronger preferences)

Item number	Mean rank	Item number	Mean rank	Item number	Mean rank
9	2.69	3	3.87	1	7.46
5	3.10	12	5.10	8	7.69
10	3.14	7	5.96	4	8.37
11	3.32	6	7.10	2	9.19

5 percent level. For each group we also counted the number of subjects who had claimed daily use of a radio receiver. The results are given in Table 8.

We obtained a Spearman rank correlation coefficient (corrected for ties) of 0.868 between the percentage of radio-users and the number of items significant at the 5 percent level. This correlation coefficient is significant at the 1 percent level. Significant rank correlations are also obtained if we use only the number of cells that show a 1 percent significance level; or if we use the data from Tables 2 (and 5) or 3 (and 6), individually; or if we use only the Central African groups, that is, exclude the Algerians and Moroccans.

Such relationships were not found when we took into account familiarity with machines other than radios. Nor is there any relationship within groups between the existence of common movement stereotypes and the use of radios. For example, we combined the Negro men groups of Kamina tested with the right and left hands into one group and then split this group into two subgroups of 30 radio-users and 30 subjects without radios. There was no difference in the number of significant direction-of-movement preferences between these two subgroups.

The presence of many radios in a group is probably an index of a better educational or acculturation level. That might explain the correlation ob-

Table 8. Summary of data on preferences and the use of radios

Groups	Number of subjects in the group	Number of cells in which a 5% level of significance was obtained	Subjects using radios	
			Number	Percentage
Algerian and Moroccan workers	92	18	77	84
Moroccan boys and girls	57	9	43	75
Negro boys (Luabo and Kamina)	130	20	124	95
Negro boys (Kangu)	106	18	95	90
Negro boys (Lokutu)	77	20	62	80
Negro adolescents (Matadi)	96	20	85	88
Negro men (Matadi)	87	8	63	72
Negro men (Kamina)	60	7	30	50
Negro girls (Luabo and Kamina)	101	17	82	81
Negro women (Kamina)	60	5	16	27

tained across groups. Because no such correlation exists within groups, there is not much reason to think that the specific use of radios determines the existence of common movement stereotypes. Another explanation for the absence of any correlation within groups could be the fact that radios, even if owned by individual subjects, are enjoyed by most people in the culture, even those who do not themselves own radios.

Differences among Central African groups. For each group from Zaïre we counted the number of cells in which differences were significant at the 1 percent level. Table 9 shows those numbers for subjects working with their right hands and Table 10 gives comparable data for those working

Table 9. Significances of the differences between the number of direction-of-movement preferences significant at the 1% level (subjects working with their right hands) in various Central African groups

		Negro boys: Luabo and Kamina	Negro boys: Kangu	Negro boys: Lokutu	Negro adoles-cents: Matadi	Negro men: Matadi	Negro men: Kamina	Negro girls: Luabo and Kamina
	Number of prefer-ences significant	9	5	9	10	2	0	4
Negro women: Kamina	1	.01	—	.01	.01	—	—	—
Negro girls: Luabo and Kamina	4	—	—	—	.05	—	—	
Negro men: Kamina	0	.01	.05	.01	.01	—		
Negro men; Matadi	2	.02	—	.02	.01			
Negro adoles-cents: Matadi	10	—	—	—				
Negro boys: Lokutu	9	—	—					
Negro boys: Kangu	5	—						

Table 10. Significances of the differences between the number of direction-of-movement preferences significant at the 1% level (subjects working with their nonpreferred left hands) in various Central African groups

		Negro boys: Luabo and Kamina	Negro boys: Kangu	Negro boys: Lokutu	Negro adoles-cents: Matadi	Negro men: Matadi	Negro men: Kamina	Negro girls: Luabo and Kamina
	Number of prefer-ences significant	7	8	7	6	1	0	6
Negro women: Kamina	1	.05	.01	.05	—	—	—	—
Negro girls: Luabo and Kamina	6	—	—	—	—	—	.05	
Negro men: Kamina	0	.01	.01	.01	.02	—		
Negro men: Matadi	1	.05	.01	.05	.01			
Negro adoles-cents: Matadi	6	—	—	—				
Negro boys: Lokutu	7	—	—					
Negro boys: Kangu	8	—						

with their left hands. In the body of each table are given the significance levels between all pairs of numbers of significant cells for each of the several groups of subjects. Fisher's exact two-tailed probability test (Siegel 1956) was used. Only those probabilities of 5 percent or less are shown.

These tables show there is not much difference between Negro men and women in their direction-of-movement preferences. However, Negro men at Matadi and at Kamina differ from Negro boys and adolescents, with the younger subjects exhibiting the stronger movement preferences. Negro girls appear to occupy an intermediate position: they show one significant difference with Negro men (Kamina) when working with the left hand, and another with Negro adolescents (Matadi) when using the right hand. These results are consistent with the hypothesis that there is a generation gap owing to differences in education, but with girls somewhat disadvantaged in this respect as compared with the boys.

Moving dial. Results with the moving dial (test item 13) show very few significant differences. These findings are consistent with the recommendation that moving scale indicators should be avoided whenever possible (see, for example, Chapanis 1965, pp. 44–47).

CONCLUSIONS

The data of this study show that some very dependable direction-of-movement stereotypes for displays and controls on the same plane are just as valid for North African and Central African subjects as for Western groups. In the Central African groups such stereotypes are much stronger in young educated people than in semiliterate adults, probably as a consequence of differences in the exposure of the two age groups to some of the simpler products of modern technology.

REFERENCES

Chapanis, A. *Man-machine engineering*. Belmont, California: Wadsworth, 1965.

Chapanis, A., and Gropper, B. A. The effect of the operator's handedness on some directional stereotypes in control-display relationships. *Human Factors*, 1968, **10**, 303–20.

Debrabandere, G. Left handedness and industrial efficiency. An experimental approach to the problem of stereotypes and handedness. Dissertation for a master's degree in psychology (in Dutch), Catholic University at Leuven, Leuven, 1971.

Siegel, S. *Nonparametric statistics for the behavioral sciences*. New York: McGraw-Hill, 1956.

Verhaegen, P., and Ntumba, A. Note on the frequency of left-handedness in African children. *Journal of Educational Psychology*, 1964, **55**, 89–90.

NINE

Overcoming the Language Barrier with Foreign Workers

FLORENT J. A. VOETS

Among the problems facing a firm employing foreign workers are that the customs and language of the new country are strange, and the workers are new to their professional tasks. Several measures can assist in alleviating these problems, for example, existing materials may be translated into the language of the foreign workers, interpreters may be used, or executives may master the foreigner's language. In most cases, however, it is better if the foreign workers acquire a basic knowledge of the new country's language and customs as soon as possible. Doing so makes for better communication and a greater degree of acceptance by their colleagues, and establishes them as persons to be taken seriously.

When teaching foreign employees, the practical aim should be to equip them with a readily accessible vocabulary so that they can function within the work situation. Our method differs in a number of ways from the approaches usually chosen, and its distinguishing features are:

1. The objective is to teach a vocational language in such a way that it, together with professional skill, will lead to higher and improved productivity. The scope of the course is, therefore, limited to the everyday language spoken within the company, which entails teaching technical and common terms within the most frequently encountered contexts and meanings (Fig. 1).
2. Apart from the spoken language, attention is also paid to specific gestures and actions, often radically different from those to which the person is accustomed, which are required for professional skill.

Florent J. A. Voets, Raadgevend Bureau Ir. B. W. Berenschot, Churchilllaan 11, Utrecht 2500, The Netherlands.

3. The course material consists of a list of expressions, turns of phrase and words used within the company. Words containing sounds which foreigners find difficult to reproduce are eliminated. Complex sentences are simplified. Particular attention is paid to words and phrases that are used in alternative ways, but in modified forms.

4. The method is geared to rapid and efficient acquisition of practical language skills. The lessons consist of a varied use of words in short standard sentences. Some basic principles used throughout the course are: a graded series of steps of increasing difficulty, repetition, training in speaking and listening, practical correlation with the known working situation, and constant checking and feedback. Audio-visual aids are used to increase the effectiveness of the method. Color slides show objects and actions and teach sounds separately from the "word image." Recording tapes enable the correct pronunciation to be checked by the pupils themselves and by the instructor.

5. The system also covers the training of the teachers in conducting and supervising educational conversation. The foremen are familiarized with the system and prepared to assist the pupils in putting their newly acquired knowledge to immediate practical use.

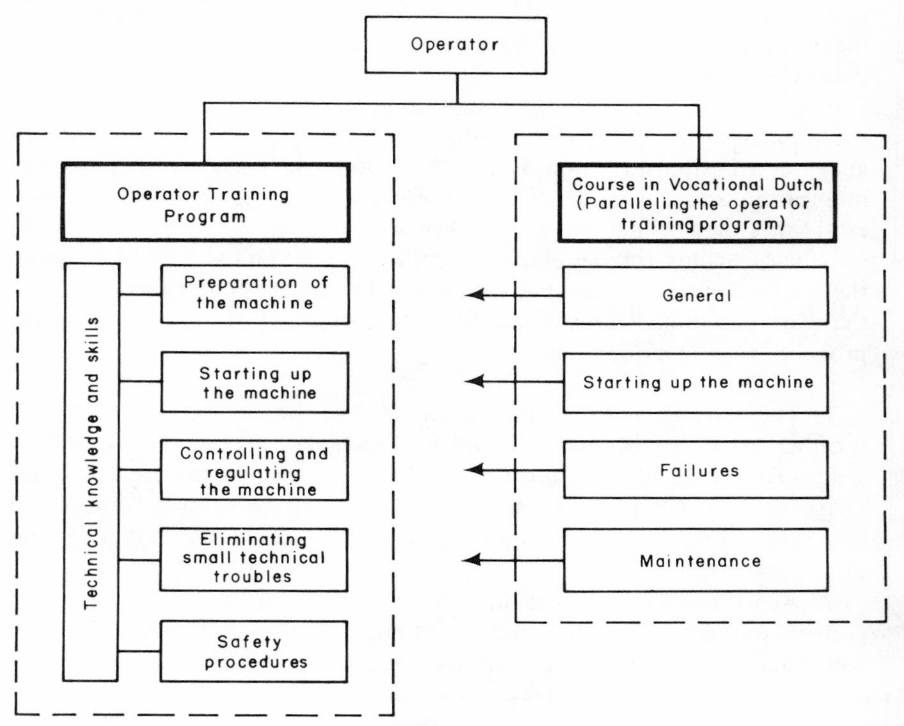

Fig. 1. An example of a typical operator-training program.

THE DEVELOPMENT OF A COURSE IN VOCATIONAL DUTCH

The course was developed by following the steps shown in overall capsule form in Figure 2. The several steps are discussed in more detail below.

GATHERING INFORMATION

The first step involved the collection of course material from interviews with workers and company personnel and from the analysis of source documents (Fig. 3). During the interviews foremen and operators whose work brings them into close contact with Spanish workers were asked to summarize the guidelines they give and the questions which the workers regularly ask. A list of reaction options was also drawn up. The overriding criteria in all this were direct relevance to the work to be done and to the language as it is used in the factory. The list of words and sentences which we assembled was then subjected to a final check by operators, foremen, instructors, and middle management.

ASSIMILATING THE INFORMATION

The assembled material then went through a number of processes before evolving as the final subject matter of the course (Fig. 4). Material was screened and classified into the most common sentence constructions and

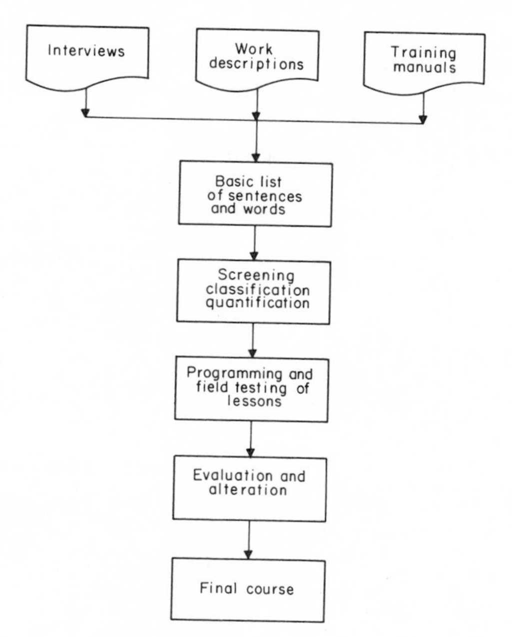

Fig. 2. Steps in the development of a course.

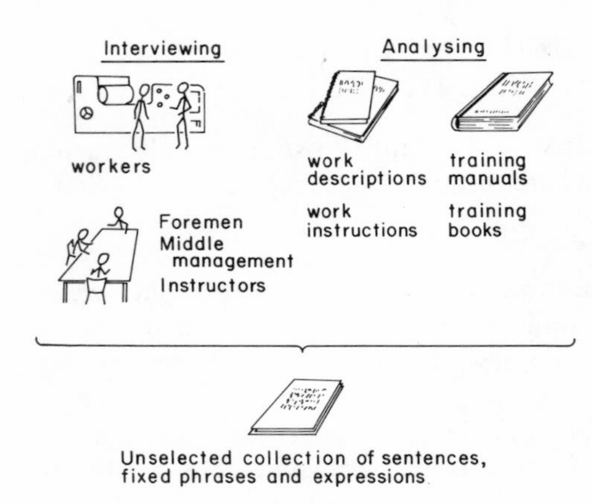

Fig. 3. Collection of information for the course.

was screened for the most current words. Sentences and words were structured according to the modifications most commonly applied to them, for example, conjugations and declensions.

PROGRAMMING THE COURSE

The material was divided up and arranged into suitable instruction units consisting of five new elements to be learned and five repetitions per unit (Fig. 5). Each instruction unit was given a methodical form composed of four parts:

1. introduction by the instructor or interpreter;
2. training by tape and slides, together with an exercise book;
3. training by assorted exercises;
4. conversation practice with the instructor.

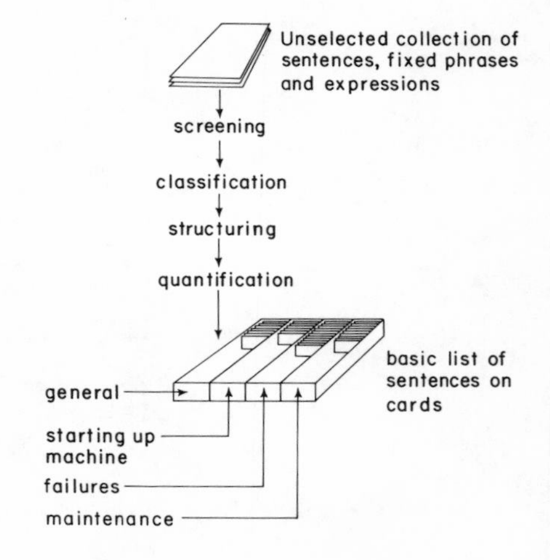

Fig. 4. Refinement of the information.

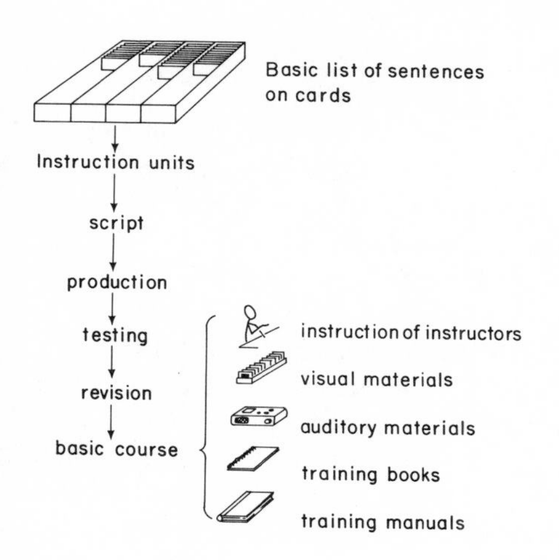

Fig. 5. Programming the course.

THE COURSE

Each course contains 25 to 50 lessons, each lasting about 60 minutes. The lessons are given 4 to 8 times a week and consist of about 15 minutes of oral introduction repeating the preceding lessons and preparing the new one. Then follows about 15 minutes of audio-visual instruction using a tape recorder and a slide projector. The lesson closes with approximately 15 minutes of oral assimilation of the lesson by means of short conversations between trainees and instructor and among the trainees themselves. At the end of each week the trainees receive an illustrated text book containing a summary of the subject matter and exercises for self-instruction.

EDUCATIONAL APPROACH

LANGUAGE AS A MEANS OF COMMUNICATION

The Dutch language is an instrumental skill which the foreigner requires to be able to communicate observations arising out of his work situation while he is working. We view the work situation as a complex one containing many variables:

- different means for communication, for example, via a colleague, by making use of gestures and facial expressions, via an interpreter, or in writing;
- different fields of perception, for example, visual, auditory, or tactual;
- different mental processes within the participants, depending on their experiences and their expectations.

When we compared practical situations with those obtaining in theory we found that the latter were very incomplete.

ANALYSIS OF LANGUAGE PATTERNS

People often think that learning a language is simply a matter of learning words, idioms, and verbs. This view does not attach sufficient importance to the structures and situations in which these linguistic elements come into play. To be able to analyze, classify, and quantify it is helpful to visualize the structure of the sentences to be learned.

Although the sentence patterns of work instructions and orders used on the shop floor vary widely, it is possible to reduce them to a single structure. Discussion of these problems with the foremen will enable them to make their usage more understandable for foreigners. Not, as so often happens, by speaking broken Dutch with the workers but by using a correct, although simplified and less varied, form of the language. When speaking broken Dutch, people generally use language patterns that are totally opposite to those of their own language but which they believe, almost invariably incorrectly, to be the structure of the foreigner's language.

PSYCHOLOGY OF INFORMATION

When drawing up a lesson we try to teach the trainees as many elements as they can absorb within the time limit of the lesson. Each lesson contains some eight or nine elements, but since most of the trainees already know some Dutch, they only receive about six or seven new items of information per lesson. If too much information is presented to them, they become confused and begin mixing up certain words and using incorrect patterns.

EVALUATION AND REVISION

Each lesson is tried out a few times with different trainees. The instructor notes down the problems and mistakes made by trainees and discusses his findings with them. The following problems and mistakes have been encountered:

- *Too much information per lesson.*
- *Necessity of unlearning such childish constructions as*, "here no paper, you look for it." Although Dutchmen use this kind of language to try to get their meaning across to foreigners, it is often extremely difficult to unlearn this kind of construction, especially for people who have already lived a long time in this country.
- *Dissimilarities with constructions in the mother tongue.* An example of this is the verb "to be" which is either left unsaid in Spanish or is not tied to a specific position:

In Dutch		*In Spanish*
"the cigarette is ready"	=	"the cigarette ready"
	=	"the cigarette ready is"

- *Differences in speed of learning.* There are short breaks of between 3 and 6 seconds duration on the tape to give the trainee the opportunity to speak. Some wait to hear what the rapid learners are saying before they speak themselves. This problem can be solved by isolating the trainees in separate cubicles. One difficulty with this solution is that the trainees like to sit together.
- *Pronunciation errors.* These are difficult to improve by using slides and tape. Our experience shows that personal instruction is required after the audio-visual training to make substantial gains in pronunciation.
- *Errors of contrast.* Opposites such as "open/closed" and "left/right" are very often interchanged. It is possible to achieve real improvement in the elimination of these errors by presenting only one of a pair in one lesson and the other in the succeeding lesson.
- *Errors of sound.* Trainees often have problems with successive words with the same initial letter. Examples are turning a switch "on" or "off" or sentences such as "*t*ry *t*o answer *t*he *t*elephone." A further difficulty is that the terminal sound of a word is often either missed altogether or incompletely pronounced.
- *Reading not effective for learning.* Although trainees improve their pronunciation by reading written texts, the overall learning result is negative. In reading written texts, the trainees do not react freely and pay more attention to the pronunciation of the words than to the contents of the text.

CONCLUSION

This paper describes a course for teaching a vocational language to foreigners quickly and correctly, so that the foreign workers can perform their tasks in accordance with the demands of the firm, and report in Dutch to inspectors or colleagues on the nature and the quality of their work.

Although our work is only in an initial phase, our experience suggests that the effectiveness of this kind of course is largely determined by or dependent on preselection of the trainees, restriction of the course to one function and one nationality, coordination between the language course and the job training program, and the quality of training for the job.

The course only shows results when used in combination with a job training program; it cannot be used on its own. Our prime objective is to give the foreign worker a thorough grounding to get him started, so that he can acquire more knowledge within the work situation due to a greater ability to communicate with others. A further aim is to achieve practical results as soon as possible.

TEN

Verbal Factors in Human Engineering: Some Cultural and Psychological Data

H. WALLACE SINAIKO

An important and neglected area of human factors engineering has to do with the problems encountered when complex equipment designed for use in one country is transferred to another. Complex equipment in this context means things like airplanes, autos, radios, and helicopters. Simpler devices that are more easily operated and maintained are not the concern of this paper because they usually present fewer problems in foreign settings. People can learn how to handle simple things. Complex equipment, however, makes special demands on men who use it or maintain it. Learning to fly, or to drive, or to repair an airplane is always difficult, even for people who are technically sophisticated. Learners need to be taught and they need to read books and instructions. In the case of complex equipment, even experienced repairmen use instructions or manuals. When equipment has been developed in one country and transferred to another, new sets of problems arise. Some of these problems are associated with cultural and psychological variables that affect learning to operate and maintain equipment. The variables that I want to talk about are verbal skills: reading, writing, and talking, and the use of words to instruct and assist people who use equipment.

This paper defines problems and provides some solutions. It is based on real situations involving the transfer of equipment, and it draws heavily on the results of field investigations and experiments. The countries involved are the United States and Vietnam, with some minor examples from Africa and the Middle East. Although I shall have nothing to say directly

H. Wallace Sinaiko, The Smithsonian Institution, Washington, D.C.

about European nations, the problems of cross-national transfer are not country-specific. Therefore, what follows about the United States and Vietnam should be of general value. Moreover, the research techniques that I shall describe are applicable to any country.

In this paper I summarize findings extracted from several technical reports and journal articles that are the result of collaborative efforts by Preston Abbott, American Institutes for Research; Richard Brislin, University of Hawaii; George Guthrie, Pennsylvania State University; George Klare, Ohio University; Lawrence Stolurow, State University of New York at Stony Brook; and myself.

APPROACH

Two sources of information were used to collect the data. First were the experiences of many people in both military and civilian agencies who have worked with foreign nationals, primarily in the developing nations. These experiences were obtained from approximately 100 interviews with people who had extensive experience in the Philippines, China, Korea, Afghanistan, Turkey, Pakistan, India, Nigeria, and Thailand. About half of the interviews were conducted in Korea and Vietnam. Most of what we learned in this phase was anecdotal in nature and is subject, of course, to modification by later, more systematic research. Discussions were not limited to military equipment or military operations. Informants included many industry-based people with responsibility for computers, jet transports, heavy construction equipment, and communication system design and installation. Some of the most valuable information came from UNICEF, the United Nations agency that is responsible for truck maintenance and repair in some of the most remote and least technologically developed countries of the world.

A second source of information was derived from three types of experiments: (1) laboratory studies of the work of language translators, including readability measurements using several hundred subject-readers; (2) field experiments requiring technicians to repair helicopters using different versions of language aids, that is, manuals; and (3) the psychological assessment of nonverbal aptitudes of Vietnamese and American subjects with a battery of tests.

FINDINGS OF THE INTERVIEW STUDY

Transferring equipment between nations is an extremely complicated process. Traditionally, however, we have assumed that men can always learn to use and maintain what is at first unfamiliar to them. That assumption seems to be substantiated by what you see if you sit in the lounge of any international airport. British VC-10s and Tridents, French Caravelles,

Russian IL 62s, and American 707s and 747s are being flown and maintained by Dutchmen, Germans, Italians, Malaysians, Indians, Roumanians, Japanese—indeed, by people from all over the world. Still, the cost of training men to work in culturally or linguistically unfamiliar technologies is very high. It takes about one year, for example, to teach a Vietnamese officer enough fundamental English so that he can begin flight training; it takes about as much time to train an airman to read and understand English so that he can begin to learn helicopter technology.

THE CAPACITY TO ABSORB TECHNICAL TRAINING

As far as we know, the psychological capacity to absorb technical training is not limited by culture (Sinaiko, Guthrie, & Abbott 1969). There is abundant evidence that the Vietnamese, and probably any other people in the world, can learn to operate and maintain highly sophisticated equipment. Our informants could think of no inherent limits to the absorption of technical information and the acquisition of new skills. For example, the Boeing company reports that Vietnamese pilots and maintenance men trained in regular courses in Seattle perform as well as their counterpart specialists who work for U.S. airlines. Other industrial contractors cited experiences with skilled heavy equipment operators whose origins were the rice paddies of Vietnam or Thailand only a few years earlier. In organizing new national airlines, Pan American World Airways has trained jet pilots and mechanics from among men who were less than a generation removed from camel drivers in Afghanistan, Iran, or Pakistan. The IBM World Trade Corporation reports similar experiences in Nigeria, where previously unskilled men were trained to handle the installation and maintenance of computers. An almost forgotten U.S. Air Force demonstration illustrates how far people from less technologically developed cultures can be taken in a short time. In the early 1950s a small group of Turkish laborers who were illiterate in their own language were sent to USAF schools in America, where, after language and technical training, they qualified as jet engine maintenance men. At least two of the men later became managers of engine repair centers in Turkey.

Such dramatic training results are, however, not easily come by. In almost every case cited, training had to be preceded by rigorous preparation in the English language. To Americans, some foreign nationals appear to acquire new knowledge slowly and their attrition rates in many courses are high. These problems may, however, be reflections more of the difficulties of working in an unfamiliar language than of any inherent learning deficiency. Finally, the most important determining factor in human learning is the motivation to learn. Given proper training and motivation, well-designed equipment, and the proper aids for using it, there seem to be no cultural barriers to anyone becoming a skilled operator or maintenance man.

CULTURAL DIFFERENCES IN LEARNING

The way people approach learning appears to differ markedly among cultures. Learning in Vietnam, and elsewhere in the Orient, is largely accomplished through imitation, memorization, and rote. This is true of theoretical or conceptual material as well as of more practical skills. There is the tendency to believe that having learned the words one has also acquired understanding and knowledge. Careful attention needs to be paid to training and to testing the effects of that training in conceptual or abstract material. Because of their strong need to "save face," Asian students rarely ask questions. To do so implies ignorance. These attitudes suggest the need for developing new approaches to technical training in Asian countries.

ATTITUDES TOWARD WORK

In some parts of the world, there are strong biases against tasks or occupations associated with "dirty hands." These attitudes are also reflected in lower average pay scales for people in such occupations. One reason for the scarcity of professional engineers among the educated Vietnamese is that engineering is considered a low-status occupation. Liberal arts, languages, and the humanities are the popular courses of study in their universities. In addition, top-level managers in some countries may be appointed because of political influence or family status rather than because of their experience or knowledge. In technical enterprises, such managers simply do not perform effectively. Their lack of expertise, together with their "no dirty hands" attitudes, makes for inadequate supervision of maintenance operations.

DIFFERENCES IN BODY SIZE AND STRENGTH

Differences in body size and strength, dealt with more extensively elsewhere in this volume, are important cultural variables with obvious implications for equipment design. As part of a larger study, Guthrie, Brislin, and Sinaiko (1970) administered a simple test of static grip strength, using a hand dynamometer, to 82 Vietnamese airmen and a smaller control group of 18 U.S. Army soldiers. The results were startling: there was almost no overlap in the two distributions. More specifically, the 75th percentile of the Vietnamese group corresponded to about the 10th percentile of the Americans. The fact that Vietnamese are much smaller than Westerners not only affects their ability to operate heavy equipment, but it may increase the number of men required to do certain maintenance tasks. Equipment has to be scaled to the physical size and strength of the people who will use it.

PERCEPTION OF TIME AND FUTURE ORIENTATION

A striking characteristic of many non-Westerners is that they appear to have no concept of preventive maintenance. This is true not only for equipment but for the human body as well: preventive medicine and dentistry are unknown in some parts of the world. In like manner, vehicles are used until they break down. Once vehicles have broken down, however, repair reflects great ingenuity, skill, and improvisation. Where comparisons are possible, military equipment is repaired as efficiently by Vietnamese as by Americans. In fact, some comparisons have yielded evidence of greater skill on the part of Vietnamese technicians. Still, preventive maintenance needs much more emphasis in training for both operators and managers.

These observations have implications for the selection of equipment to be transferred from America to certain other countries: where possible, the equipment selected should be minimally dependent on elaborate preventive maintenance. In the design of new equipment, it may often be a better strategy to accept somewhat lower performance in a system if maintenance demands can thereby be greatly reduced. Since industrial experience in Vietnam has shown that training in preventive methods can be vastly improved, new approaches to such training are also needed. In developing such new approaches, it is useful to consider the experiences and unconventional maintenance practices of organizations operating in the developing nations. For example, technical training should be limited to teaching a few skills instead of aiming at the production of generalists. UNICEF experience has shown that poor operator behavior accounts for more vehicle breakdowns than any other cause, implying that maintenance problems may be reduced as much by training operators as by training maintenance men. Speed governors and automatic transmissions have also been used to improve this situation, and special mobile maintenance teams have been employed to provide preventive measures on regular schedules. Complex diagnostic aids are used even less by Vietnamese than by American technicians. Attention should be paid to developing special job aids, for example, checklists, simple diagrams, and graphics, that can be used both to train new men and to assist experienced maintenance people.

SUMMARY

There are some exciting prospects for solutions to maintenance problems, although most of these solutions have been developed for Americans. Recent research sponsored by the U.S. Department of Defense has produced new materials and methods for training in troubleshooting and repair for a wide range of complex equipment. This research has shown that programs of instruction are highly dependent on documents, leading to the design of new technical manuals that are radically different and based on known principles of human learning. Not only do these new documents provide faster and better training, but they become long-term aids to experienced technicians in the field. Experiments have demonstrated the value of these approaches to maintenance of a wide range of complex equipment, for example, C-141A aircraft, UH-1F helicopters, and other sophisticated systems. There is some uncertainty, however, about whether these methods will be as effective with non-American user populations.

SOME GENERAL LANGUAGE ISSUES

By "language issues" I mean all the verbal material related to learning (e.g., textbooks); operations (e.g., maps or verbal instructions); and maintenance (e.g., field manuals). There are two general solutions to the problem of giving documents written in one language to people who read another. First, as mentioned earlier, one can expect the recipients to learn to read and to speak a second or unfamiliar language. This is an old and commonly followed solution. Most scientists who are neither British, American, nor Canadian have had to learn English as a second language. Without making a judgment about individual experiences, I assert that learning English is slow and expensive and that use of a second language requires constant practice and reinforcement. A reader will have greater or lesser skill in English, or any second language, depending on his need to use it. Many scientists are conversationally proficient at the level of general social interaction. But are they able to read and understand the English of electronic theory or automotive engine repair? Americans who have lived in the United Kingdom have an appreciation of still another problem, the real but subtle lexical differences between native speakers of English in the two countries: "boot" in the United Kingdom means "trunk" in the United States, "bonnet" means "hood," "wing" means "fender," and so on.

A second solution is translation. This is also a very costly activity, and even when done well it has special problems. First, direct translation may not even be possible between some languages. Vietnamese is a rich language when it deals with literature, the humanities, and philosophical subjects. It is impoverished, however, when dealing with technology. Traditionally, the Vietnamese have had to use French or Chinese terms when they treat technical subjects. Another problem of translation is that the people needed to do it, skilled professional translators, are often in very short supply. This is particularly true of technical translators. It is a dangerous mistake to assume that any bilingual person can do translations at all or that he can handle unfamiliar subjects. In view of these problems, it is not surprising to find many technical translations not done well (see Fig. 1).

SPEED OF TRANSLATION

Surprisingly, very little research has been done on translation. Our studies show that good translations are done very slowly: there seems to be an upper limit of about 400 to 450 words an hour among good translators (Sinaiko & Brislin 1970, 1973). Curiously, these figures are the same for translations between many languages, for example, English-to-French, Vietnamese-to-English, and Russian-to-English. In spite of anecdotal claims to the contrary, we have never seen 1,000 words per hour produced when technical subjects are involved (see Fig. 2). Incidentally, translation from Vietnamese to English is about 10 percent faster than English to Vietnamese, probably because the back translator does not have to spend time trying to understand the meaning of highly technical terms that have already been put in simpler phrasing by the original translator.

> Fall down the rear seat back on the seat, and insert two stays at the both side of the floor on the top of the seat back into the small hole of the steel brackets on the back of the seat cushion.

Fig. 1. Manufacturer's translation from Japanese in the owner's manual for the Datsun 1964 station wagon.

UNFAMILIAR TERMS

Another problem of translation is the way bilinguals are forced to handle unfamiliar terms. In our observations, based on many thousands of words translated into Vietnamese, translators did one of several things: (a) they simply used the English term, perhaps transliterated to accommodate the target language, for example, "piston" became "pittong"; (b) they substituted a term from a third language, such as French or Chinese; (c) they coined a new term; (d) they used available words to "talk around" the untranslatable word or to describe it functionally. An example of the last is "tachometer," for which there is no Vietnamese equivalent. Our translators would typically substitute such a phrase as "rotation measuring machine," which could be translated into Vietnamese.

ERRORS IN TRANSLATIONS

For every 100 technical English words translated into Vietnamese, our data show that about five words or combinations of words were in error. Whether this error rate is satisfactory or tolerable is a matter of practical judgment. In a later section, however, I give some findings on the impact of translation quality on performance. Here are some types of errors that we observed:

- Words, whole phrases, or even sentences, were omitted.
- Ambiguous synonyms were substituted for directly equivalent and available terms.
- Words or phrases became garbled.
- Meaning was changed through the choice of wrong words or terms.

There was, incidentally, no correlation between translator speed and accuracy, although this may have been an artifact of the very restricted range of performance of our highly selected translator-subjects.

IMPROVING THE QUALITY OF TRANSLATIONS

Several steps can be taken to improve translations. First, the original language of the material to be translated can often be better written. This may be viewed as a first-order translation, that is, from poor English to good English. Sinaiko and Brislin (1970) suggest several rules for writing more easily translatable English, among which are:

- Do not write in long sentences, that is, in sentences of more than 16 words.
- Do not use complex noun phrases, for example, "organizational maintenance activities."

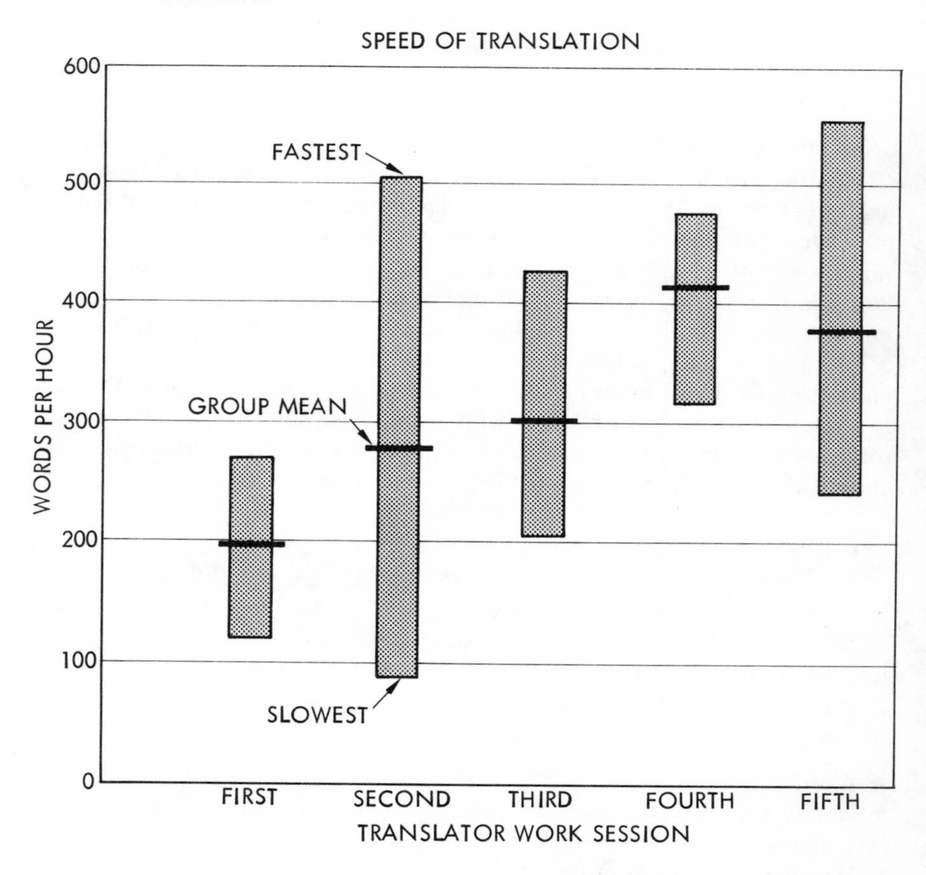

Fig. 2. Speed of translation from English to Vietnamese (Sinaiko & Brislin 1970).

- Avoid adverbs and prepositions that indicate degree, for example, "reasonably probable," or "beyond."
- Avoid abbreviations, for example, "Landing Gear Cont."

Incidentally, an advantage of better English is that it also benefits native readers of that language.

Another approach to better translation is to provide the translators with aids such as bilingual glossaries and technical dictionaries. This is not a trivial recommendation. Major translation services, such as those in Mannheim, Federal Republic of Germany, and in Luxembourg, have invested many years of linguistic research effort in the development of terminological bureaus for the creation of technical terms and lists.

Finally, the control of translation quality can be improved. In the past, quality has generally been ignored or, at best, handled subjectively; that is, a translator's work may be sampled by a bilingual editor or supervisor who simply compares the original material with its translation to see if they are "similar." Experimentally derived evidence shows that such comparisons are not a satisfactory means of judging quality and that, in fact, they may be completely erroneous. We have seen the following discrepancies occur: (a) a group of Vietnamese technicians who said they did not like a translation actually performed very accurately when using the material; (b) a group of Vietnamese technical advisors pronounced another translation of the same material as "not bad" (subjective opinion), but this translation resulted in poor maintenance performance by technicians working on the same task.

TRANSLATION QUALITY AND WORK PERFORMANCE

One of our studies posed the question: "How do translations of varying quality or accuracy affect the work of technicians who are required to use them?" Recently we ran a series of experiments addressing this issue (Sinaiko & Brislin 1970, 1973). We selected a very demanding maintenance task involving the UH-1H helicopter. Because of its complexity—the task included twelve sequential steps having to do with the adjustment of a power turbine generator—even experienced technicians had to use the maintenance manual to do the job (see Fig. 3).

PROCEDURE

Our subjects were 72 Vietnamese Air Force helicopter mechanics who had recently been trained in their specialty. As control subjects we used a smaller group of U.S. Army helicopter technicians with identical training and experience. Our experiment was straightforward. Men from the Vietnamese group were randomly selected and assigned to three-man teams. Each team was required to perform the twelve-step maintenance

5-391. ADJUSTMENT - POWER TURBINE GOVERNOR RPM CONTPOLS.

a. Be sure collective pitch control system rigging has been completed.

b. Lock collective pitch control stick in full down position and adjust droop compensator control tube (2, figure 5-77) to align center of bolt hole in aft arm cf torque tube (4) approximately level with top of support bracket (11). Due to shimming, manufacturers tolerance, etc., variation of 0.250-inch from top of support bracket is possible and acceptable. (See detail B.)

c. Set cam adjustment (15) to middle of slot. (See detail D.)

d. Move collective pitch control stick to full up position and lock.

e. Adjust control rod (12) attached to cam bellcrank so that approximately 0.25-inch of cam slot is visible below cambox housing for T53-L-9, -9A, -11, and -11B engines; 0.38-inch for T53-L-13 engines.

This is a nominal setting and is subject to change, if necessary, in following steps.

f. Check installation of governor control lever (17) as nearly at 90 degree angle to stop arm as serration alignment permits. (Refer to paragraph 5-386 step a.)

g. Adjust upper governor stop screw to 0.250-inch for T53-L-9 and -9A engine; 0.210-inch for T53-L-11, -11B, and -13 engines, measured from inner side of mounting boss. (See detail C.) Remove and discard lead seal on lockwire, if existing.

Note

Never shorten either stop screw on governor to less than 0.060 inch length from inner side of boss.

h. Disconnect actuator shaft from governor control lever (17) by removing bolt.

i. Electrically position actuator shaft to approximate midpoint of stroke.

(1) If actuator with two adjusting screws is installed, turn both positive stop adjusting screws to obtain maximum stroke (see detail E). Reduce stroke by turning each screw ten full turns away from maximum adjustment to obtain actuator nominal position.

(2) If actuator with single adjusting screw is installed it is not necessary to adjust positive stop screw to obtain nominal position. Positive stops can be adjusted, if necessary, for travel of 0.500 inch to 1.75 inch without change in nominal position.

Note

One full turn of the adjusting screw will cause a change in both the retract and extend position of .032 inch. (See detail E.)

Note

Set actuator travel to: 1.38-inch for T53-L-9, -9A engines; 1.25-inch for T53-L-11 series engines; 1.20-inch for T53-L-13 engine.

j. Fully retract actuator shaft by holding GOV RPM switch to INCR. Move collective stick to full up position.

k. Reinstall bolt connecting actuator to governor control lever, adjusting actuator shaft rod-end to obtain 0.010 inch clearance between governor stop arm and upper stop screw, measured with a feeler gage. (See detail C.) If necessary, reposition control lever one serration on governor shaft to accomplish this adjustment while keeping safe thread engagement of rod-ends.

Fig. 3. A portion of the maintenance instructions from U.S. Army's "UH-1H Technical Manual," TM-55-1520-210-20, Section 5-391.

task using one of four qualities of translated material: (a) a high quality translation into Vietnamese produced under our control; (b) a lesser quality translation produced by a "free lance" translator over whom we had no quality control; (c) a poor quality translation purchased through a commercial service; or, (d) the original English language material.

MEASURES

We measured two things: speed of performance and precision. Measurement of the latter consisted of observations made by expert instructors, who noted, for each of the twelve steps in the task, one of the following: "no error," "minor error" (the men were able to correct their mistake), or "major error" (the men could not continue the task without an observer's intervention). For each language condition there were six three-man teams of subjects.

FINDINGS

The results of the experiment are shown in Table 1. Two interesting conclusions can be drawn from the data. First, it appears to make a substantial difference whether or not one works in his native tongue or in a second language. Note that the Vietnamese who were forced to use English, although they had learned that language reasonably well and had been trained as technicians in that language, did much less well than the American controls who also worked with English instructions. Second, the quality of translated material had a strong influence on performance: the best translations produced performance nearly equal to that of the Americans, while the succeedingly poorer translations resulted in correspondingly poorer performances and much more frequent occurrences of major errors. Although this type of testing is more expensive and time-consuming than other methods of evaluating translations, such as the use of reading tests or the analysis of back-translations, we feel it is the most valid way to assess translation quality. So far as we know, this is the only experiment that has tested and shown a relationship between translation accuracy and performance.

Incidentally, speed measures in this experiment were meaningless because of the inability of some subjects to do the job at all with the poorest translation.

TRANSLATION BY COMPUTER

Some recent experiments (Klare, Sinaiko, & Stolurow 1972; Sinaiko & Klare 1972, 1973) illustrate a different, but related, aspect of ergonomics and culture. These studies came about as a result of recent developments in the field of computational linguistics, the field that is concerned with applying computer technology to the analysis of natural language. One such application is the automatic translation of languages. Although translation by computer has been in development for at least fifteen years, and there is at least one operational system that regularly translates Russian to English, very little has been done to measure the quality of machine translations. How good are machine translations?

About two years ago an English-to-Vietnamese machine translation system became available. The system was attractive, primarily because it

Table 1. Accuracy of maintenance operations by Vietnamese and U.S. mechanics using four different versions of a maintenance manual[a]

Translation used	Errors		
	None	Minor	Major
Vietnamese: supervised translation	73.1%	21.3%	5.6%
Vietnamese: free-lance translation	40.3	55.5	4.2
Vietnamese: commercial translation	11.0	52.0	37.0
English (VNAF subjects)	40.7	39.6	20.6
English (U.S. Army subjects)	73.2	26.7	0.0

[a] Sinaiko & Brislin 1970.

promised to break up the serious backlog of technical manuals awaiting translation by conventional, human means. What was unknown, however, was anything about the quality of the translations.

PROCEDURES

Our approach consisted of measuring the readability or comprehensibility of the translations on the assumption that readability is correlated with quality. Our methods, incidentally, are general: they are applicable to most written material, whether in translation or in its original form.

Specifically we constructed and administered knowledge or reading tests; used a little-known readability measure known as the "cloze procedure" (Klare et al. 1972; Taylor 1953); administered comprehensibility rating scales; and measured reading rate or speed.

The translated material consisted of an instructional text, a standard U.S. Air Force text on instrument flying, and a Navy operational order called "Casualty Control." Untranslated samples of 500-word passages, in the original English, formed the control material. Both American and Vietnamese subjects, the latter having been trained in English, took tests using this control material. The second condition consisted of the same material translated into Vietnamese by the best human translators available. In fact, these were some of the same men who provided the highest quality translations used in the helicopter experiment. The third and fourth conditions consisted of the material translated into Vietnamese by computer. In one case we used raw or unedited translations; in the other, post-edited computer translations.

The subjects were 88 U.S. Air Force student pilots who were familiar with the instrument flying material, 172 Vietnamese student pilots who were familiar with the English version of the material, 57 U.S. Navy officer candidates who had not seen the text but who had studied the subject of casualty control, and 141 Vietnamese Navy officer candidates who had training similar to that of the American U.S. Navy officer candidates. Subjects from the Vietnamese groups were assigned randomly to one of the three translation conditions. Americans, of course, read only the English text in each case.

FINDINGS

Results of the readability experiments showed that translations by highly skilled bilinguals were superior to the best machine translations. For example, using reading comprehension tests as a criterion, mean scores for the Vietnamese Air Force subjects were 66 percent on the human translations, 57 percent on the edited machine translations, and 41 percent on the unedited machine translations. An unexpected finding was that the Vietnamese subjects made their highest scores (69 percent) with the English material. Furthermore, the more technically complex the material, the greater the loss in comprehension on the computer translations. For example, reading test means were 62 percent for the human translation of the most difficult material and 18 percent for the unedited machine translation of the same passage. Another measure, reading speed, did not differentiate among the three versions of Vietnamese, although subjects reading English as a second language were much slower. Finally, as the text became more technically complicated, reading speeds dropped correspondingly.

The cloze procedure is a very stringent test of reading comprehension because it tests the understanding of structure as well as meaning. It also differentiated the three translations and in the same order. English comprehension scores were lowest (20 percent) for Vietnamese subjects reading the original English versus 53 percent on the human translations, 46 percent on the edited machine translations, and 33 percent on the unedited machine translations. Finally, subjective ratings of comprehensibility were highest for the post-edited computer translations and the English text.

CONCLUSION

In conclusion, two general things can be said about the quality of translations done by computer. First, the best post-edited machine translations were surprisingly good, but not as readable as the best that could be done by skilled human translators. Second, technical material became relatively less readable in translation as it increased in complexity, particularly for the unedited machine versions.

CULTURAL DIFFERENCES IN ABILITY

While related to language variables, ability and aptitude measures should be examined in their own right, because they may be different among people of different cultural backgrounds. During the course of the experiment on the maintenance of helicopters we observed an apparent problem among the Vietnamese technicians: when they came to one of the technical drawings, the subjects seemed to have difficulty interpreting the figure (see Fig. 4). This led one of my colleagues, George Guthrie, to propose a new experiment that would try to assess some of the nonverbal abilities of the Vietnamese. It was Guthrie's hypothesis that perhaps our Vietnamese subjects did not have the same perceptual abilities for interpreting graphic representations as did Americans. To test this notion we administered two types of tests to groups of U.S. Army and Vietnamese helicopter technicians (Guthrie, Sinaiko, & Brislin 1971). First, we gave a battery of eleven tests that measured various spatial and numerical abilities (see Table 2). Eighty-two Vietnamese and 31 Americans took the tests. Second, we administered an experimental test, one developed by Hudson in his

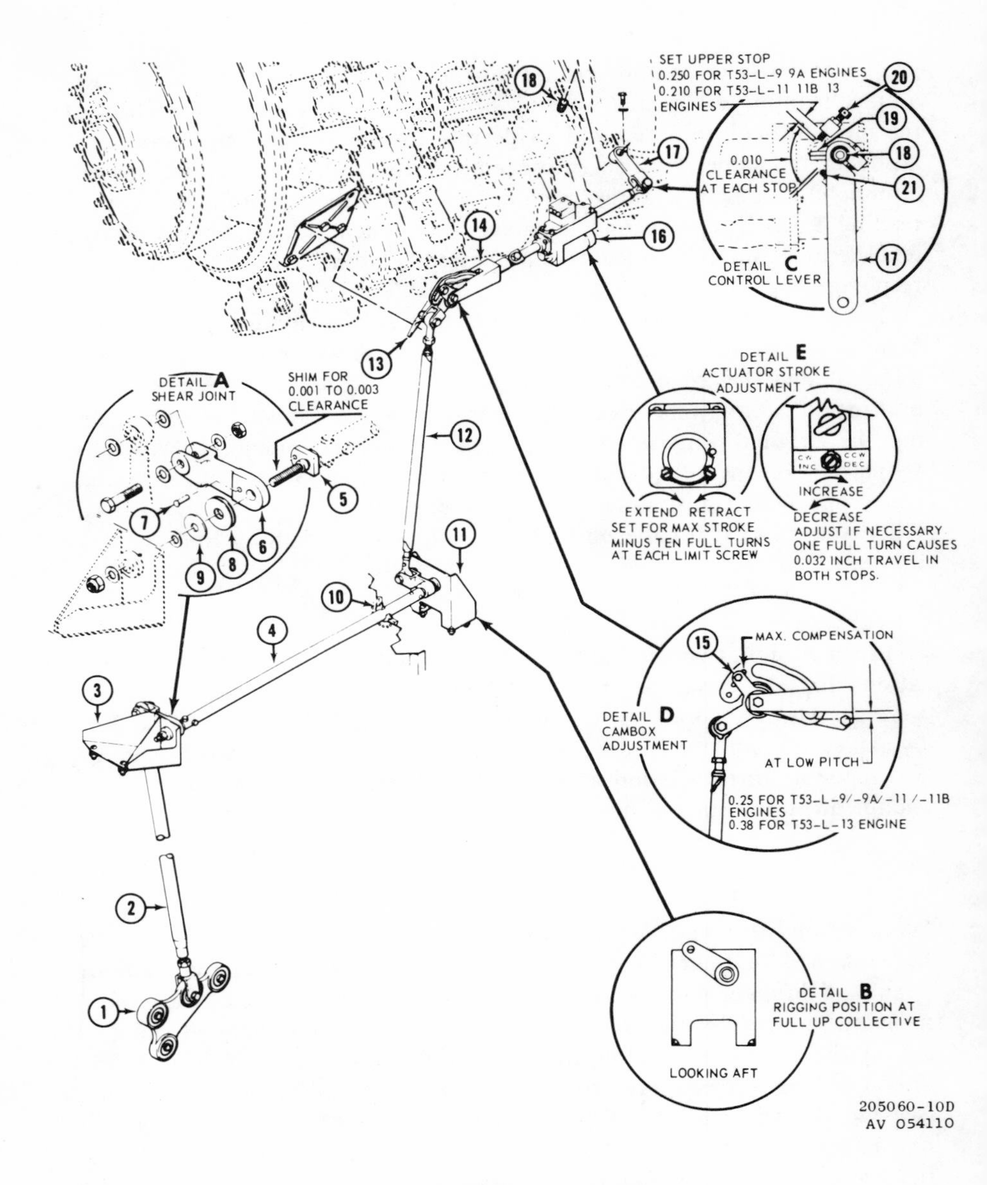

Power turbine governor rpm controls

Fig. 4. A technical illustration that accompanies the maintenance instructions in Figure 3.

Table 2. Description of the test battery (from French, Ekstrom, & Price 1963) administered to Vietnamese and American helicopter technicians

Test	Factor tested	Time limit (min.)
1. Addition	Numerical ability	4
2. Division	Numerical ability	4
3. Card rotation	Spatial orientation	8
4. Cube comparison	Spatial orientation	6
5. Paper-folding	Visualization	6
6. Number comparison	Perceptual speed	3
7. Identical pictures	Perceptual speed	3
8. Length estimation	Length estimation	6
9. Nearer point	Length estimation	4
10. Hidden patterns	Flexibility of closure	4
11. Tool knowledge	Mechanical knowledge	10

work with Africans (Hudson 1960, 1968), to assess ability to interpret two-dimensional graphic figures. Eighteen American and 18 Vietnamese subjects took the latter test. All tests were of the so-called "culture-fair" or nonverbal type. Figure 5 shows the six items from Hudson's test. Subjects simply indicate the target of the hunter's spear.

FINDINGS

Although detailed results of the testing appear in the journal literature (Guthrie, Brislin, & Sinaiko 1971), some observations are worth repeating here. The Vietnamese subjects made significantly higher mean scores on two of the tests (numerical addition and flexibility of closure) than did the American controls. On five tests (spatial orientation, perceptual speed, and mechanical or tool knowledge) the American subjects scored significantly higher. On the remaining five tests in the battery there were no differences between the Americans and Vietnamese. The test of graphic interpretation, Hudson's figures, yielded results consistent with the above. Out of 108 responses by Vietnamese subjects, 37 percent were in error, using a Western criterion, while only 9 percent of those by the American controls were so judged.

POSSIBLE ARTIFACTS IN THE TEST DATA

A possible error of test construction, perpetrated unknowingly by psychological test developers, may account for some of these measured differences in ability when, in fact, such differences are not real. Figure 6 shows two test items from the "tool knowledge" material. The two items on the left are stimulus items. The subject is supposed to pick the one alternative, A, B, or C, that goes with the stimulus item. The lower item, showing a tire wrench as the stimulus object, was answered correctly by only 38 percent of the Vietnamese subjects, while 98 percent of the Americans knew the correct answer. The top item was answered correctly by 95 percent and 100 percent of the two groups, respectively. Inspection of the two sets of figures shows a probable reason for these differences: the artist, drawing for a primarily Western audience, did not consider scale inconsistencies, which to an American viewer would be no problem. This problem is dealt with more thoroughly in the paper by Wyndham in this volume.

CONCLUSIONS

Our conclusions and recommendations from the measurement of nonverbal abilities were:

- Illustrative material might present a great deal of difficulty to non-Western readers, more than is often believed to be the case.
- There is a growing body of data, confirmed by our tests, that people of different cultures employ different conventions in portraying and interpreting three dimensions represented in two-dimensional space.

Fig. 5. Hudson's test.

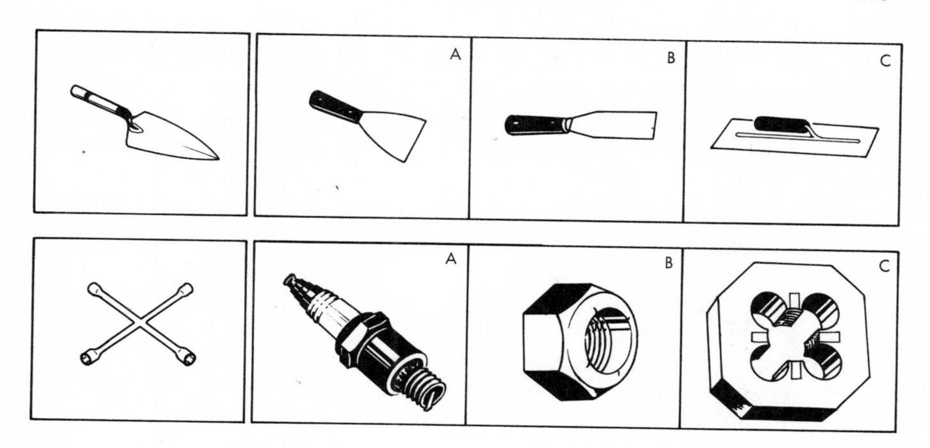

Fig. 6. Two test items from the tool-knowledge test.

- If the results obtained on our mechanical knowledge tests are valid, Vietnamese or other mechanics might have difficulties in selecting and using correct tools.
- Deficiencies in perceiving spatial patterns can probably be eliminated through training, as has been done successfully in some non-Western groups.

SUMMARY

The main points to be drawn from this paper are the following:

1. There appear to be few cultural barriers to the operation and maintenance of complex equipment. But when equipment is designed for multinational use, human factors engineers must take into account certain important cultural variables: the way people learn, their attitudes toward maintenance and repair, their stature and strength, and the way they perceive graphic representation.

2. Language factors are extremely important, and, unfortunately, they have been neglected. There are essentially three ways to handle language problems: (a) teach the user your language but expect this to be slow and expensive, (b) translate your language into that of the user, but expect this to be slow and costly and subject to error, or (c) eliminate dependence on language by minimizing the need for reading skills. There are some promising automatic translation techniques, but they are far from operationally useful at the present time. Finally, translation quality is related to work performance, strikingly so in some of our results. Better translations produce better work, and vice versa.

3. There may be ability or aptitude differences among cultural groups. They should be recognized and measured before machines developed for one part of the world are delivered to another.

4. Some general suggestions for the human factors scientist who finds himself working in a cross-cultural setting are:

(a) Good human factors engineering practices are always important, but in cross-cultural situations they require an added emphasis. Do not assume that design or training principles that have proved useful in one country are equally valid in another. This is most important as cultural distances widen.

(b) There is useful information in the accumulated knowledge and observations of cultural anthropology, and human factors specialists should become familiar with it.

(c) Similarly, there is a wealth of nonscientific experience among business men and others who have worked in settings initially foreign to them. Although such information should not be substituted for data obtained under carefully controlled conditions, it is, nevertheless, valuable.

(d) There is an urgent need for more experimentation in cross-cultural settings.

REFERENCES

French, J. W., Ekstrom, R. B., and Price, L. A. *Kit of reference tests for cognitive factors*. Princeton, New Jersey; Educational Testing Service, 1963.

Guthrie, G. M., Brislin, R., and Sinaiko, H. W. Some aptitudes and abilities of Vietnamese technicians: implications for training. Paper P-659. Arlington, Virginia: Institute for Defense Analyses, November 1970.

Guthrie, G. M., Sinaiko, H. W., and Brislin, R. Nonverbal abilities of Americans and Vietnamese. *Journal of Social Psychology*, 1971, **84**, 183–90.

Hudson, W. Pictorial depth perception in sub-cultural groups in Africa. *Journal of Social Psychology*, 1960, **52**, 183–208.

Hudson, W. The study of the problem of pictorial perception among unacculturated groups. *International Journal of Psychology*, 1968, **2**, 89–107.

Klare, G. R., Sinaiko, H. W., and Stolurow, L. M. The cloze procedure: a convenient readability test for training materials and translations. *International Review of Applied Psychology*, 1972, **21**, 77–106.

Sinaiko, H. W., and Brislin, R. W. Experiments in language translation: technical English-to-Vietnamese. Research Paper P-634. Arlington, Virginia: Institute for Defense Analyses, July 1970.

Sinaiko, H. W., and Brislin, R. W. Evaluating language translations: experiments on three assessment methods. *Journal of Applied Psychology*, 1973, **57**, 328–34.

Sinaiko, H. W., Guthrie, G. M., and Abbott, P. S. Operating and maintaining complex military equipment: a study of training problems in the Republic of Vietnam. Research Paper P-501. Arlington, Virginia: Institute for Defense Analyses, July 1969.

Sinaiko, H. W., and Klare, G. R. Further experiments in language translation: readability of computer translations. *ITL*, 1972, **15**, 1–29.

Sinaiko, H. W., and Klare, G. R. Further experiments in language translation: a second evaluation of the readability of computer translations. *ITL*, 1973, **19**, 29–52.

Taylor, W. L. "Cloze procedure": A new tool for measuring readability. *Journalism Quarterly*, 1953, **30**, 415–33.

ELEVEN

Multinational Psychological Research in Telephony

R. B. ARCHBOLD

The Consultative Committee on International Telegraphy and Telephony (CCITT) with headquarters in Geneva, Switzerland, is one of the two technical branches of the International Telecommunications Union (ITU). The other branch is the Consultative Committee on International Radio (CCIR). The CCITT studies technical and operating questions relating to international telephony and telegraphy and prepares appropriate recommendations on common standards which member organizations can use in designing equipment, systems, and procedures.

The members of the CCITT are the telephone administrations of the 140 countries which are members of the ITU (e.g., the German Bundespost), 42 recognized private operating agencies (e.g., the American Telephone and Telegraph Company), 85 scientific or industrial organizations (e.g., Kakusai Denshin Denwa Co., Ltd., in Japan), and 16 international organizations (e.g., the International Standards Organisation).

The work of the CCITT is carried out formally by one of nineteen study groups through the study of various questions. The actual work program is entrusted by the Study Group to one of its working parties.

TASKS OF THE HUMAN FACTORS WORKING PARTY

One of these working parties (Working Party II/5, that is, Party 5 in Study Group II) was specifically set up in April 1966 as a Human Factors working party to study what problems might arise when telephone subscribers were eventually able to dial their own international calls. These problems included: how to standardize the printing of telephone numbers on letter-

R. B. Archbold, Post Office Research Department, Dollis Hill, London NW2 7DT, England.

heads for international understanding; whether the subscriber would be confused by unfamiliar audible tones for "ring," "engaged," etc.; what mistakes the subscriber might make in dialling the longer digit sequences involved in international numbers, and what are the best formats to recommend for telephone numbers to keep errors to a minimum; and what symbols and layouts would be universally acceptable for expanding the pushbutton telephone pad from ten to sixteen buttons.

GENERAL METHODOLOGY

It was recognized from the start that in view of national, cultural and other differences, it was important to standardize the methods of study. Data differences could then more easily be attributed to national differences rather than to artifacts of different experimental and analytic methods.

For each study, therefore, a "rapporteur," sophisticated in experimental design, sampling theory and data analysis, was chosen to propose a definitive method. After this proposal was accepted by the working party, it was then followed rigorously by the countries participating in the study; each country was of course responsible for the translation of instructions. In this way the same stimuli, the same procedures, and the same data analysis were used by all participants. Results were sent to the rapporteur, who summarized them in a final report for consideration by the working party at its next meeting.

SOME REPRESENTATIVE STUDIES

AUDIBLE TONE STUDY

The first major study carried out by the working party was concerned with whether national subscribers could correctly distinguish between ringing, engaged (busy), and other tones heard from another country. In particular, an experiment was designed to determine how much confusion might be caused with present tones, what characteristics of a tone can be varied without generating confusion, and what the design requirements should be for future new tones. I describe in some detail the methodology used in this study as an example of the kind of procedures generally followed in the studies discussed later in this paper.

Four parameters of tones were varied (see Table 1) to produce a set of fifty-four signals which cover the interesting parts of the four-dimensional space. To these were added ten tones which were representative either of types of existing tones or of tones to be introduced shortly in certain countries (see Table 2). Finally, the tests included eight so-called "ambiguous" or unreal telephone tones. All tones were recorded on tape and were presented to subjects in a partially counterbalanced order within the same experiment. Subjects were asked to say whether they thought each tone was "ringing," "engaged," or "neither ringing nor engaged." If a subject

Table 1. Parameter values used in constructing the set of 54 test tones

Signal parameters	Values
Modulation	M_1 = 0 Hz M_2 = 16 Hz M_3 = 50 Hz
Frequency	f_1 = 425 Hz f_2 = 620 Hz
Repetition period	P_1 = 0.3 sec P_2 = 1.0 sec P_3 = 6.0 sec
Ratio $= \frac{\text{Signal on period}}{\text{Repetition period}}$	R_1 = 0.1 R_2 = 0.33 R_3 = 0.5

Table 2. Characteristics of the real telephone tones used in this experiment

Identification of tone	Frequency in Hz	Repetition period in seconds	On period	Modulation in Hz
Ringing tones				
1. American Precise Tone Plan (PTP) ringing tone*	440 + 480	6.0	2.0 sec	0
2. German ringing tone*	450	5.0	1.0 sec	0
3. Irish and British (UK) ringing tone	400 + 450	3.0	400–400 msec	0
4. Japanese ringing tone	400	3.0	1.0 sec	16 2/3
5. Approximate ringing tone used in Czechoslovakia, France, Holland, Pakistan, Poland, Switzerland, and the USSR	425	5.0	1.0 sec	0
Busy tones				
6. American "Line busy"	600	1.0	500 msec	120
7. American PTP "Line busy"*	480 + 620	1.0	500 msec	0
8. American PTP "Reorder (trunks busy or congestion)"*	480 + 620	0.5	200 msec	0
Special tones				
9. French transfer tone	(The actual French tone was taped and used)			
10. CCITT special tone	950/1400/1800	2.0	320–320–320 msec	0

*These tones were not yet in use at the time this paper was prepared.

responded "neither," he was then asked whether it sounded more like a ringing tone or more like an engaged (busy) tone. In this way the responses to each tone were classified as ringing or busy and as real or unreal. That is, the percentage of responses judged "ringing" (Table 3) is based on the number of subjects who judged a tone to be ringing, plus those who judged it to be neither ringing nor engaged, but more like a ringing tone. The percentage of "unreal" judgments is based on the number of subjects who said that the tone was neither ringing nor engaged, regardless of whether they then said that the tone sounded more like a ringing tone or more like a busy tone.

Tests were conducted in the following nine countries with the numbers of subjects shown in parentheses: Australia (32), France (225), Germany (35), Holland (16), Japan (32), Spain (32), Sweden (31), the United Kingdom (60), and the United States (32). In most countries the subjects were mixed engineering, clerical, and administrative people, largely from telephone administrations or the telephone industry. In the U.K. tests, however, sixty Cambridge housewives were used. If we can assume that these several groups of subjects were more accustomed to their own country's national signals rather than the much broader range of international signals, then each group's background of experience with its own familiar telephone signals is part of its "culture."

Illustrative of the type of information obtained in this study are the confusion data for the ten tones representative of existing tones (see Table 3). Since the data clearly show that certain existing tones are difficult to interpret, the various telephone administrations have been encouraged to explore this problem further through actual field tests. Results on the set of fifty-four signals indicated that, of the four tone characteristics studied, repetition period and modulation seem to be more important than frequency and on-off ratio.

While it is costly and confusing to the public to have telephone tones changed, the results obtained have been useful in suggesting standards to be followed in the design of new systems and in providing instructions to subscribers to help discriminate existing tones.

Further experiments were conducted in some countries to compare the efficacy of different forms of instruction, both written and tape recorded, in simply generalized and precisely detailed forms, together with graphical representations in both simple and complex forms.

DEFINITION OF SYMBOLS

Symbols play an important international role in several fields including communication and transportation. The working party has a continuing program under way to provide symbols for procedures and equipment. In view of the conflicts between countries that were found in the use of symbols prior to international dialling, the working party began by obtaining

Table 3. Percent of experimental subjects in each of nine countries judging each tone in the left-hand column as a "ringing" tone

Tone*	Australia	France	Federal Republic of Germany	Holland	Japan	Spain	Sweden	United Kingdom	United States
Ringing tones									
1. American PTP	97	76	97	100	100	100	100	85	97
2. German	38	84	100	100	75	100	100	37	81
3. Irish and British	100	47	74	69	50	37	63	100	55
4. Japanese	100	87	100	94	100	97	100	90	91
5. Czech, French, Dutch, etc.	44	81	94	100	69	100	100	42	87
Busy tones									
6. American "Line busy"	26	28	4	0	0	0	0	58	0
7. American PTP "Line busy"	19	21	4	0	0	0	7	33	0
8. American PTP "Reorder"	6	14	0	0	9	0	0	30	3
Special tones									
9. French transfer	3	27	4	13	34	0	3	18	6
10. CCITT special	31	46	29	38	56	48	33	53	44

*These tones are the same as those in Table 2.

agreement on a definition of four classes of symbols for national and international numbers, as follows:

Diallable symbols. A diallable symbol is a symbol which is to be dialled. It appears on a telephone set to designate either a finger-hole of a dial or a pushbutton of a key set. These symbols can be digits, letters, or other signs.

Procedural symbols. A procedural symbol is a symbol which tells the subscriber how to dial. Such symbols should not appear in a finger-hole or on a pushbutton because they are not to be dialled.

Information symbols. An information symbol is a symbol associated with the subscriber number describing special features of the subscriber telephone service. For example, the symbol Ꝋ,[1] where used, indicates that the subscriber has an answering device attached to his telephone. Such symbols are not to be dialled and therefore should not appear in a finger-hole or on a pushbutton, nor can such symbols be procedural in instructing the subscriber how to dial.

Spacing symbols. Spacing symbols are symbols which are used solely to separate parts of a telephone number from each other. They cannot be diallable, procedural, or information symbols.

These definitions are embodied in a recommendation that has been provisionally accepted by the CCITT and that will materially add to the uniformity of number presentation on letterheads sent to other countries and in directories and dialling instructions for foreign visitors.

SYMBOLS FOR EXTRA BUTTONS ON KEYSETS

The working party addressed itself next to the problem of standardizing the designations of buttons 11 and 12 for a 12-button telephone set, and later buttons 13–16 for a 16-button set (see Fig. 1).

As an aid in reaching agreement on a set of standard symbols for this purpose, it was useful to define some desirable properties of such symbols.

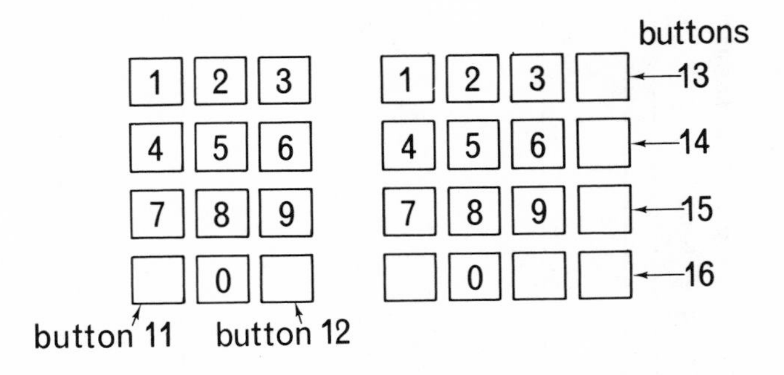

Fig. 1. A 12-button telephone set showing buttons 11 and 12 (*left*); a 16-button keyset showing buttons 13–16 (*right*).

[1] *Editor's note*: This symbol cannot be reproduced exactly in type. It resembles a reversed Q.

Accordingly, it was agreed that some desirable properties to be considered when selecting diallable symbols are as follows:

1. *Distinctness from other diallable symbols.* As used here, "distinctness" refers to dissimilarity from other symbols when compared with them visually or aurally. This dissimilarity should be evident in a low probability of confusion with other symbols under degraded perceptual conditions.
   - The symbols should be visually distinct in their designated form as well as in typewritten, handwritten or printed form, including variations which might occur in each.
   - The symbols should be aurally distinct from one another when they are named in at least the official languages of the ITU, English, French, and Spanish.
2. *Widely known name.* The name of the symbol should be as widely known as possible and should be constant over as wide a range of the population as possible.
3. *Reproducible.* The symbol should be easily reproducible in handwritten and typewritten form.
4. *CCITT—ISO compatible.* The symbol should be a member of the CCITT Alphabet No. 5 and the International Standards Organisation (ISO) standard code for information interchange.
5. *Made up of a single character.* The symbol should not be composed of more than one individually valid symbol, nor should more than one key operation on a typewriter, for example, be required to produce it.
6. *Abstract.* The symbol should not already have intrinsic meaning resulting from other specialized usage.
7. *Immediately recognizable as a diallable character.* The symbol should not be one which is used for procedural or information purposes.

The criteria listed above for such symbols were considered in selecting candidates among various alternative sets. In practice, it is not possible to satisfy all criteria, and some compromises must be (and were) made. In addition, experiments were conducted to measure dialling speed and accuracy, auditory confusions in acoustic transmission, and visual confusions in handwriting symbols.

The symbols chosen for the 11th and 12th buttons, * and #,[2] could not be alphabetic letters because of possible confusion with letters already provided on buttons 2–9 at the time when the symbols for the 11th and 12th buttons (* and #) were chosen. At that time the telephone numbers in many countries were still composed of a combination of letters and digits. Alphabetic letters were therefore excluded from consideration, since no letter can be on more than one button.

The more recent experiments on choosing symbols for buttons 13–16 were able to consider letters, since most countries are well on the way to converting to all-digit dialling. Letters, of course, were fairly obvious choices because they satisfied so many of the desirable criteria listed above.

[2]*Editor's note*: These symbols cannot be reproduced exactly in type (see Fig. 2 for their exact shapes and dimensions).

Specifically, criteria 2, 3, 4, and 5 together rule out most alternative symbol sets. Given that letters were a likely choice, it was necessary to select a suitable set of four letters from the alphabet. Accordingly, two sets of experiments were conducted.

In one set designed to study the legibility of handwritten characters, strings of seven characters composed of a random ordering of two random digits, four random letters, and either * or # (see footnote 2) were generated and printed by a computer. Subjects in each of the eight participating countries were asked to copy 50 such strings by hand, and the handwritten version was then given to a keypunch operator to transcribe back into computer readable form. The result was compared by the computer to the original material, and data were presented as the proportion of times each handwritten character was mispunched and what errors were made.

The whole experiment was performed separately for upper and lower case letters, and confusion data were published for all letters for each participating country. The results were, of course, directly comparable with one another because the materials and procedure were identical and, consequently, the differences in confusion patterns among countries were attributable to the different, culturally determined, handwriting habits.

The other set of experiments used similar combinations of letters, digits, and the symbols * and # (see footnote 2), but in an acoustic confusability setting. Each of the seven countries participating in this study used a similar experimental design, utilizing speakers and listeners native to that country. The experiment alternated subjects as speakers and listeners and measured the errors on each character under two circuit conditions chosen to simulate interesting, real world circumstances. Again, the set of experiments was so carried out that the differences in confusion patterns were attributable to cultural (and language) differences exclusively, for example, the pronunciation of the letter *z* as *zee* in the United States or as *zed* in the United Kingdom giving rise to confusion patterns with other letters and digits.

The results from both series of experiments served to permit the selection of A, B, C, and D for buttons 13–16 with the additional requirement that a phonetic equivalent would be provided for each.

A QUESTIONNAIRE INTERVIEW

Another area of concern to the various telephone administrations concerns the sources of errors made in placing telephone calls, the effectiveness of various information sources, and the problems and annoyances experienced by the telephone user. The solutions for those problems that affect international calls need to be implemented internationally and to achieve this, agreement is required on the nature and extent of the problems. Accordingly, two questionnaires were constructed, one for a user of an international circuit, and one for a foreign traveler. These questionnaires were designed to be administered in identical form in each country, so that the results would be directly comparable. The concept of "identical form" includes identical questions (translated as needed), identical evaluative choices (or items of a rating scale), and identical rules for administration. A large-scale program for use of these questionnaires has already begun, and preliminary tests in a number of countries have shown the questionnaires to be usable and highly informative.

LAYOUT OF PUSHBUTTON DIAL

Another series of recent studies, conducted jointly and internationally, relates to the layout of a 12-button and 16-button telephone set. A 4 × 3, 12-button layout had previously been agreed on as the international standard, but special design needs had raised the question of an alternative standard, viz., a 6 × 2 array. At about the same time, it became necessary to standardize a 16-button set, and a decision had to be made whether it would be necessary to separate perceptually an added column of four buttons from the three already in the 4 × 3 set.

A single comprehensive experiment was designed to provide insight on both issues. The experiment included a comparison of 4 × 3 with 4 × 4 and with 6 × 2 arrays (see Fig. 2) in terms of dialling speed and error rate. The identical experimental design was used by six countries. Although identical pushbutton sets were to have been used in each country, problems of scheduling were such that this was not completely adhered to. In addition, some aspects of data recording presented severe problems for some of the participants. As a result the overall data analysis was not as neat as it might have been, but meaningful and important results were nevertheless obtained. Specifically, it was concluded that the 6 × 2 array was slightly inferior to the 4 × 3 array (but nevertheless usable), and that no indications were obtained of a need for perceptually separating the extra column of a 4 × 4 telephone in that errors and dialling speeds were substantially the same for the 4 × 3 and 4 × 4 arrangements. These conclusions were generally consistent among different countries.

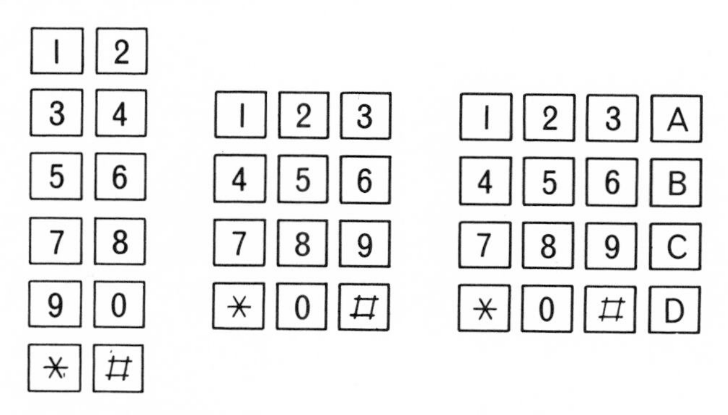

Fig. 2. The 6 × 2 (*left*), 4 × 3 (*center*) and 4 × 4 (*right*) keysets tested in this experiment.

CONCLUSIONS

The main conclusion to be drawn from these studies is that problems of international telephony involve human factors that are broader than, and that could not be discovered from, studies conducted in any one country. At the same time, these studies show that worldwide standardization in human factors work is possible and can lead to successful and worth-while agreements. It also appears that a specific organization may be needed to sponsor the work and define common problems, as in the part played in telecommunications by the CCITT Human Factors Working Party. The need for the latter is supported to some extent by consideration of the fact that human factors experiments are, in many instances, fairly expensive to conduct.

It is worth noting, however, that frequently the dedication of a few people stimulates interest and progress even within the specific organization. In less application-oriented situations than the one we have discussed, it may be possible that when a commitment has been made by potential participants, an energetic coordinator or relatively small team could reach successful conclusions in many instances.

ACKNOWLEDGMENTS

The author wishes to thank the Director of Research of the United Kingdom Post Office for permission to present this paper, and also to accord his gratitude to Dr. J. E. Karlin and Dr. M. S. Schoeffler of the Bell Telephone Laboratories for their support.

TWELVE

Human Factors in International Keyboard Arrangement

LEWIS F. HANES

Guidelines have recently been formulated to help in the design of business equipment keyboards for effective operator use (Alden, Daniels, & Kanarick 1970; Seibel 1972). These guidelines recommend assigning the primary alpha and/or numeric key cluster(s) to a central location and distributing the remaining keys with their specialized symbols, controls, and functions around the periphery of the central cluster(s). The alpha and numeric key placements usually conform to historical precedents and accepted national standards. The peripheral placement of keys is guided by such principles as frequency of use, importance, functional relationships, sequence of use, and historical precedent.

These guidelines have been used with reasonable success to configure keyboard equipment the operation of which is limited to a single and well-defined cultural group. But are these same guidelines valid when a keyboard is designed for users representing different nationalities, languages, and cultures? Or must the designer consider additional, sometimes conflicting, guidelines?

SOME PROBLEMS IN DESIGNING KEYBOARDS FOR INTERNATIONAL USE

The problems in designing business equipment keyboards for more than one nation or cultural region are complex. The designer must consider the application requirements in each region, historical precedents, and accepted and proposed national standard arrangements. These considera-

Lewis F. Hanes, Research and Development, The National Cash Register Company, Dayton, Ohio; now at Westinghouse Electric Corporation, Pittsburgh, Pennsylvania.

tions may be complicated by cultural and language variations, by international standards, and by manufacturing pressures to minimize the number of different keyboard arrangements for a given product.

Keyboards range in size from the 13-key array found on simple adding machines to the more than 220-key arrays required to generate in excess of 3,200 characters on some Kanji language teleprinters. A keyboard may have only a ten-key numeric cluster and a few function keys. Or, it may contain control keys, function keys, a separate full-key matrix numeric array, and a cluster containing alpha, numeric, punctuation, and symbol keys.

The keyboard characters may be in any one of such diverse languages as Arabic, English, Greek, Hebrew, Kanji, Katakana, or Russian. Even countries that share a common basic alphabet differ in their requirements for characters. For example, some require special characters to handle regional peculiarities, diacritical marks to provide proper phonetic emphasis, and unique alpha character arrangements based on historical precedent.

To add to the problems already enumerated, a number of international and national standards recommend keyboard arrangements for some types of machines, but the several standards do not all agree.

Manufacturers, of course, prefer to minimize the number of specialized keyboards they are required to produce for a given machine. Each additional kind of keyboard costs money for tooling, manufacture, inventory, shipping, and the preparation of specialized training and maintenance manuals.

Finally, organizations purchasing business equipment are interested in commonality among keyboards, where practical. Keyboard commonality facilitates the exchange of operators within and across national boundaries and minimizes the requirements for operator training and retraining. However, the advantages of commonality are lost if it produces significantly higher machine cost or poorer operator performance than would be possible with a specialized arrangement.

KEYBOARD ARRANGEMENT: NATIONAL

GUIDELINES

Guidelines to help the designer in arranging the keys on a keyboard (Alden et al. 1970; Seibel 1972) have developed primarily from experiences within single and well-defined national and cultural groups. They do not consider the problems associated with international operation. These national guidelines consist of at least ten rules:

1. *Determine the characters and number of keys required.* The specific alpha, numeric, symbol, punctuation, function, and control characters needed for all the intended uses of the system, and the number of keys required to contain the characters, should be determined by analyzing the machine application requirements.

2. *Arrange the keys according to their frequency of occurrence and according to operator characteristics.* Within a language, individual characters and character sequences occur with differing frequencies. Since the several fingers of the hand are not equally strong, keys should be arranged to take advantage of the strengths of certain fingers and to minimize the impact of the weaknesses of others. That is, the most frequently used characters should be located so that the stronger fingers and hand do more of the work. The arrangement of the keys should, however, allow an operator to alternate between the two hands during touch operation, since maximum speed of entry occurs with such hand alternation.

3. *Follow historical precedent.* The adage, "Do not change something unless there is a good reason," is the foremost principle in arranging the touch area of a keyboard. Since operators may have had previous experience with other keyboards, the layout should permit maximum possible transfer of experience. Most users are not willing to accept a new keyboard arrangement for a machine that performs the same basic functions as their previous equipment, unless the advantages of the new arrangement greatly outweigh the retraining problem that will result from its adoption. This guideline pertains primarily to the touch-operated region; the resistance to change is not nearly so strong for areas where visual search is normally required to locate the desired key.

4. *Follow established standards.* National standards for keyboard arrangement have been issued for some machines. Such standards should be followed unless there are good reasons to deviate.

5. *Group frequently used keys in the touch area.* In touch operation, the key position or value does not have to be verified visually. Touch-operated keys are located in the normal "home" or resting hand position, where they may be easily operated with minimum fatigue and strain. Frequently used keys should be arranged and located to facilitate touch operation. In many business equipment machines, typewriter and/or numeric groupings comprise the most frequently used keys. For that reason, they are often placed in the optimum touch area.

Function, control and symbol keys may be positioned for touch operation if they have high frequencies of use. Such keys are usually placed around the typewriter and/or numeric key cluster(s). Keys used less frequently should be placed in less accessible peripheral locations.

6. *Group common functions together.* Keys representing related functions should be grouped in close proximity to each other; for example, the keys containing the symbols for open and close parentheses should be located adjacent to each other.

7. *Group logically and according to sequence of use.* Keys should be placed in a sequence that reflects a normal and logical sequence of operation. For example, the "open parenthesis" should be placed on the key to the left of the "close parenthesis," at least in countries where the sequence of entry is left to right.

8. *Locate according to importance.* If the consequence of inadvertent key operation is great, locate the key in a peripheral location. Provide a lock on the key if greater protection is required. Keys critical to the completion of a task should normally be located according to other guidelines. Unique coding may be utilized to provide extra emphasis.

9. *Code the keys.* Key tips should be coded uniquely to facilitate the recognition of logical groupings of keys, keys of special importance, and keys that are operated with high frequency. Coding techniques include key size, shape, color, surface texture, caption style, direction of movement, and spacing.

10. *Consider all factors.* In arriving at the final arrangement, consider all relevant factors: frequency of occurrence and importance of each machine application, location and space available for the keyboard, costs, and manufacturing requirements. A trade-off study between alternatives may be necessary to define the best layout.

In principle, following the rules above should result in a keyboard providing best possible operator performance. In practice, the rules sometimes conflict. For example, following a historical precedent for the alphabetic characters (rule 3) would eliminate the possibility of grouping alpha characters according to their frequency of occurrence in the language and according to the strengths of the fingers and hands (rule 2). In other words, the guidelines cannot be followed cookbook style.

ALPHABETIC ARRANGEMENT

A great deal of controversy over keyboard arrangements has been generated by the second and third guidelines above. In particular, attempts have been made to introduce touch layouts not in agreement with previous operator experience. It is important to review this effort because of the special problems of acceptance that arise when touch arrangements for alphabetic and numeric characters deviate from those established through long usage.

The QWERTY keyboard arrangement, so called because these are the six keys in the upper left-hand part of the keyboard, has become the accepted layout in the United States and certain other countries for typewriters and for communication and data processing machines requiring alphanumeric data entry. Attempts to introduce alternative arrangements have been relatively unsuccessful for applications involving touch operation.

The Simplified Keyboard, developed by Dvorak (Dvorak, Merrick, Dealey, & Ford 1936), is supposed to be scientifically arranged on the basis of the frequency of use of letters, letter patterns and sequences in the English language. A study by Strong (1956) for the U.S. government compared performance on the Dvorak keyboard with the QWERTY arrangement. The results are often cited as a major reason for not changing the

QWERTY layout. Strong found that when experienced typists used the Simplified Keyboard, they required an average of 100 hours of retraining to achieve their original gross rate of speed. He concluded that the results of the comparison did not justify adoption of the Dvorak keyboard for use by the U.S. government. Mettler (1971) reports that similar scientifically derived arrangements have been undertaken for other languages and that the results are different in every instance, because the frequencies of the individual letters and their sequences vary in each language.

Another possible arrangement is an alphabetical sequence from A to Z, the so-called ALPHA arrangement, used on some stockmarket inquiry terminals. Results of two recent studies (Hirsch 1970; Michaels 1971) do not provide evidence that supports the selection of the ALPHA arrangement in preference to the QWERTY layout, at least for nonrandom code inputs. In the Hirsch study, subjects unskilled in typing used either an ALPHA or QWERTY keyboard to enter familiar names and some numbers. Initial performance was better on the QWERTY arrangement, and this difference was maintained after equivalent amounts of practice.

Michaels compared performance on the two keyboards with skilled, semiskilled, and unskilled typists. The input material was a list of names and addresses taken from telephone directories. He found that keying rates and work output were greater for skilled and semiskilled typists on the QWERTY keyboard. Performance on the two keyboards was essentially equal for unskilled typists. The superior performance by skilled and semiskilled typists on the standard typewriter is probably explained by their previous experience. Interference and conflicting motor responses probably occurred when these subjects used the ALPHA arrangement.

Other keyboard designs, such as chord keyboards, provide alternative configurations. Chord keyboards, which require the simultaneous depression of two or more keys, greatly increase the number of different codes that can be entered in a single movement. With 10 keys, 1,023 different codes are possible (Klemmer 1971). Conrad and Longman (1965), and Bowen and Guinness (1965) provide descriptions and experimental comparisons among chord keyboards and more traditional ones. A comprehensive review of chord keyboards and other entry devices has been prepared by Seibel (1972).

Still other keyboards of unique design have been proposed from time to time. For example, Kroemer (1972) has tested a typewriter keyboard designed to minimize unnatural, uncomfortable, and fatiguing body and arm postures. Keys on the keyboard are arranged in two hand-configured groupings—one for each hand. Keyboard sections allotted to each hand are physically separated to facilitate positioning of the fingers, and are inclined laterally to reduce muscular strain.

Why have QWERTY keyboard arrangements retained their popularity, even when some other keyboards have shown performance advantages?

Michaels (1971, p. 425) provides a succinct answer: "Probably because skill in using QWERTY is so widespread." Surveys of job skills in some American cities show that about 50 percent of the working-age population uses a typewriter for some purpose. This figure will become larger with the increased emphasis on typing in schools and the growing use of keyboard terminals for entering data into computer systems. Michaels (1971) reports that about 2.3 million typewriters are sold yearly, and there are about 45 million typewriters existing in the United States. Ancona, Garland, and Tropsa (1971) estimate that in a few years nearly two million keyboard devices will be in use in the United States alone for communication purposes and for direct computer processor entry.

Such widespread experience and skill with the QWERTY keyboard makes retraining a formidable barrier to any alternative proposal. If all keyboards were replaced with a new version, all operators would require retraining. If only part of the keyboards in an office were changed, the hiring and assignment of trained personnel would be more complicated, and restrictions on the interchange of machines during equipment malfunctions could raise significant problems.

The gradual erosion of interest in the Dvorak layout demonstrates that an alternative must have more than publicity and staunch supporters to overcome the training and acceptance obstacles. According to Mettler (1971), nonstandard arrangements have fared no better in other countries than in the United States. To be accepted, any new alpha keyboard layout must demonstrate significant advantages in performance, minimum retraining time, and an initial learning rate equivalent to or better than that possible with the QWERTY configuration. Even with such well-documented justification, new layouts may not be accepted by manufacturers and users.

NUMERIC ARRANGEMENT

There are two common digit arrangements for ten-key keyboards. The adding machine arrangement has the numbers 7, 8, and 9 on the top row; 4, 5, and 6 on the middle row; and 1, 2, and 3 on the bottom row. The touch tone telephone configuration has the 7, 8, and 9 keys interchanged with the 1, 2, and 3 keys. Other layouts have been tested (e.g., Conrad 1966; Deininger 1960), but generally discarded. Hanes and Baker (1965) concluded that experienced touch operators would have essentially equivalent performance with either configuration if they did not transfer between numbering schemes. Transfer might cause interference, reducing the speed of entry, accuracy, or both. The touch system is based on an operator's developing an almost automatic set of arm, hand, and finger movement patterns in response to the numbers to be entered on a particular keyboard. Requiring an operator to enter the same numbers on a keyboard with a different numbering arrangement would probably force him to suppress one movement pattern and develop another. There are no reports in the general literature on such transfer experiments involving skilled operators.

An experiment by Conrad and Hull (1968) found that, for initial performance by completely inexperienced subjects, the touch tone layout had a small and nonsignificant speed advantage, and a highly significant accuracy advantage compared to the adding machine arrangement.

The widespread adoption of touch tone telephones means that many people will have had previous experience on the touch tone arrangement when they come to a work situation. The impact of this previous experience on learning the configuration on adding machine keyboards is not known. Nor do we know the magnitude of the retraining problem that would arise if one layout were adopted as a universal standard, or the magnitude of the interference that might occur if both layouts existed in the same office and an operator were required to operate both.

Chord keyboards have been used in some post office letter-sorting machines, both in the United States and in the United Kingdom. Some of these machines are apparently being modified by having their chord keyboards replaced with sequential entry keyboards. Reports documenting the reasons for this change are not generally available.

The full-key matrix keyboard is another numeric arrangement in such widespread use that it has become a de facto standard, at least in the business equipment field. The number of key columns provided is usually dependent on the application.

A Belgian Post Office study (Bertelson & De Cae 1961) compared the performance of two groups of experienced operators, one group skilled on the ten-key keyboard and the other skilled on the matrix array. The results showed the ten-key unit to be faster, even when as many as 45 percent of the digits were zero. No differences in error rates were found between the two keyboards.

Full-key matrix keyboards are being provided on many machines, even though Seibel (1972), based on a review of research results, concluded that the ten-key array is preferred for skilled and unskilled operators. The retraining problem is often given as a reason that full-key units on new products are not replaced by ten-key numeric keyboards.

NATIONAL STANDARDIZATION

Keyboard standardization is in progress at both national and international levels. National standardization bodies usually develop and approve keyboard standards for their respective countries. The names and addresses of the national bodies are available from the International Standards Organisation (ISO).[1]

[1]Requests for information concerning the work of ISO should be addressed to: ISO Central Secretariat, 1, rue de Varembé, 1211 Genève 20, Switzerland.

The United States group is the American National Standards Institute (ANSI) (Ancona et al. 1971). It has issued at least three standards relating to keyboards:

1. *American Standard—Typewriter Keyboards, X4.7-1966 July 1966.* This Standard prescribes the arrangement of the 42 basic printing keys on the typewriter; the characters, upper and lower case, that appear on the keys; and the 2 shift keys and space bar. An electric typewriter keyboard is preferred, and a manual typewriter keyboard is specified as an alternate.

2. *USA Standard—10-Key Keyboard for Adding and Calculating Machines, X4.6-1966, September 1966.* This standard prescribes the arrangement of the ten numeric keys for adding and calculating machines of the ten-key type.

3. *American National Standard—Alphanumeric Keyboard Arrangements Accommodating the Character Sets of ASCII and ASCSOCR, X4.14-1971, March 1971.* The two arrangements defined are intended for typewriter-like data communications and data processing alpha-numeric keyboards implementing the character sets of the American Standard Code for Information Interchange (ASCII) and the American Standard Character Set for Optical Character Recognition (ASCSOCR).

An ANSI Standard represents a consensus of those substantially concerned with its scope and provisions. It is a guide; it does not preclude anyone from manufacturing, marketing, purchasing, or using products, processes, or procedures not conforming to the standard.

Other national standardization bodies are also involved in establishing keyboard standards. For example, the British Standards Institution (BSI) has published adding machine (BS 1909: 1963) and typewriter (BS 2481: 1961) standards (Whitfield 1971). The Japanese Industrial Standards Committee has issued Standard JIS B9509-1964, establishing a Katakana typewriter layout. Both the Canadian Standards Association and the Deutscher Normenausschuss are known to have circulated proposals for keyboard standards.

Sometimes groups other than a national standardization body may establish their own keyboard standards. For example, the United States Department of Defense (1969) has published a military standard on keyboard arrangements.

GUIDELINE EVALUATION

It is difficult to measure the success of guidelines and standards for key arrangement, since very few published studies deal with that issue. The human factors group at National Cash Register has, however, completed several proprietary investigations on a number of alternative keyboard layouts. Specifically, keyboard arrangements developed by following guidelines were compared with other layouts developed both within and outside the company. The results showed that operator performance is better on keyboards that follow guidelines than on those that do not.

KEYBOARD ARRANGEMENT: INTERNATIONAL

INTERNATIONAL KEYBOARD VARIATIONS

Keyboards for international use must consider differences between nations and cultures. At the very least, basic alphabets, such as Arabic, Greek, Hebrew, Katakana, and Roman, must be accommodated. Even within a single alphabet, special symbols and characters are required to satisfy variations peculiar to some cultures and languages. In addition, common characters within a basic alphabet may be assigned to different positions on a keyboard for historical reasons.

A comparison of international keyboards designed for one type of application, accounting, illustrates some of the differences that may occur among various countries and regions. Figure 1 shows the basic keyboard arrangement for the recently released (1972) NCR 399 accounting system. Figures 2 to 10 illustrate alternative keyboards offered as standard for 17 countries or regions. Still other variations are possible by special customer order. The key clusters in Figure 2 to 10 have been rearranged to accommodate graphics of maximum size on the key tips. The key clusters for all regions are arranged as shown in Figure 1.

Figures 11 to 14 show the keyboard arrangements for the NCR 395 accounting machine designed for four different countries, and Figure 15 shows the keyboard arrangement for the NCR 33 accounting machine designed for Turkey.

Differences among basic alphabets can be seen by comparing characters on the English (Fig. 3), Japanese Katakana (Fig. 10), Arabic (Fig. 11), Hebrew (Fig. 13), and Greek (Fig. 14) keyboards. Even within common alphabets, language differences require different keytip captions. For example, "Load" on the South African keyboard (Fig. 4) is "Laden" on the German configuration (Fig. 6), and "Charg" on the French version (Fig. 5).

Several of the figures reveal differences in character locations within the same basic alphabet. For example, the QWERTY sequence on the American keyboard (Fig. 1) is AZERTY on the French keyboard (Fig. 5), and QWERTZ on the German layout (Fig. 6). Other differences are readily apparent in the figures. Frequency of use and historical precedents are the primary reasons for the differences among countries.

The requirement for additional characters within a basic alphabet to satisfy national requirements is illustrated by the Ã, Ñ, and Õ characters found on the Spanish keyboard (Fig. 2). These characters are not required on the American version (Fig. 1).

Some symbols are required in one country, but not in another; for example, the # symbol is used on the keyboard for the United States

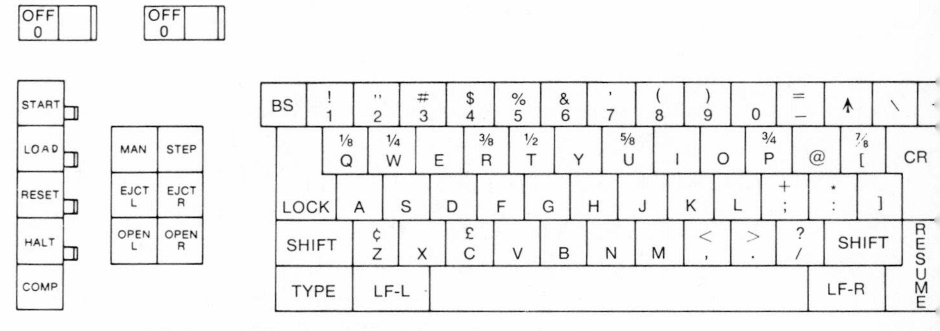

Fig. 1. NCR 399 U.S. keyboard arrangement.

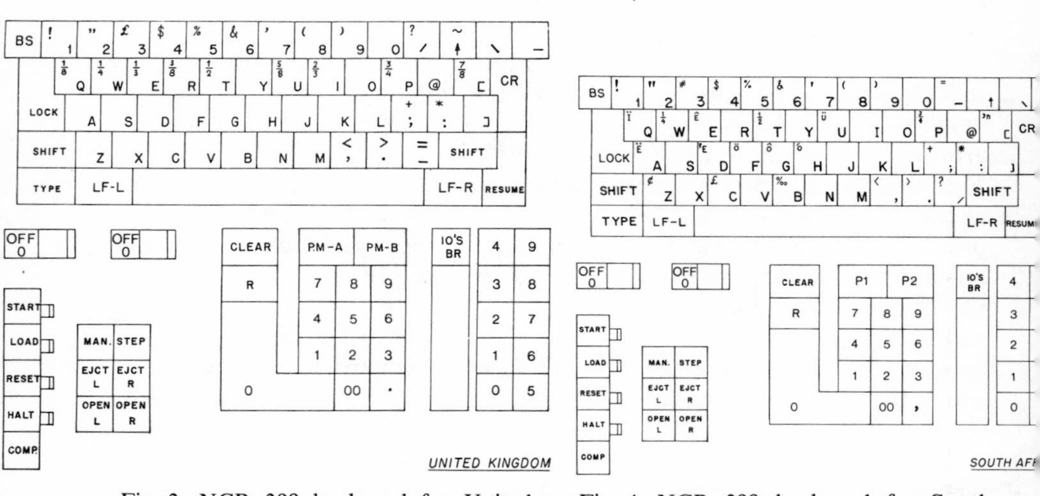

Fig. 3. NCR 399 keyboard for United Kingdom.

Fig. 4. NCR 399 keyboard for South Africa.

Fig. 7. NCR 399 keyboard for Sweden and Finland.

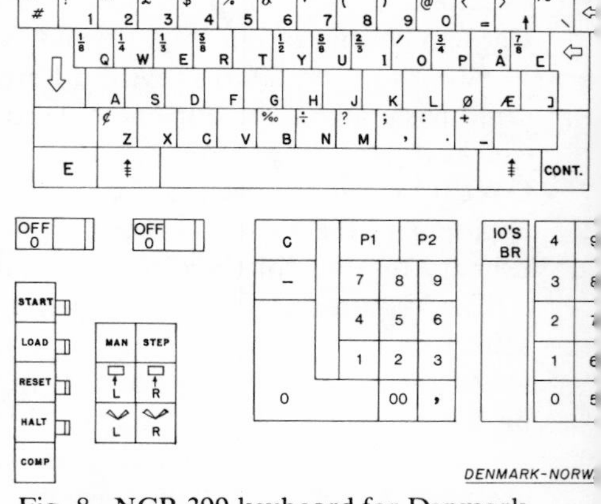

Fig. 8. NCR 399 keyboard for Denmark and Norway.

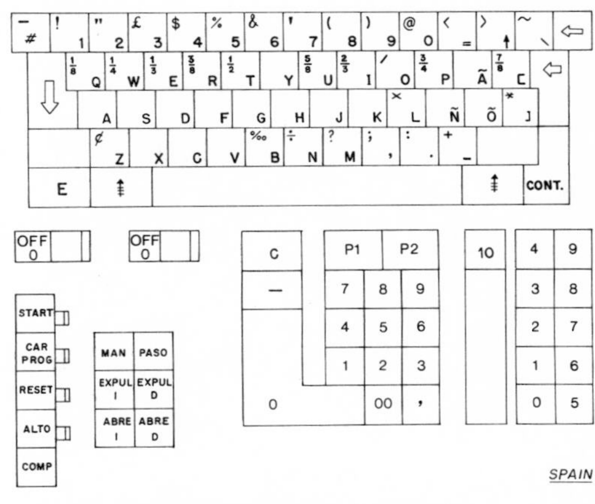

Fig. 2. NCR 399 keyboard for Spain.

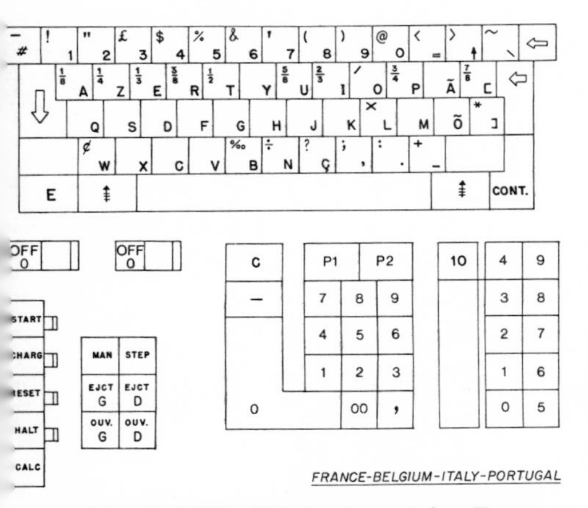

Fig. 5. NCR 399 keyboard for France, Belgium, Italy, and Portugal.

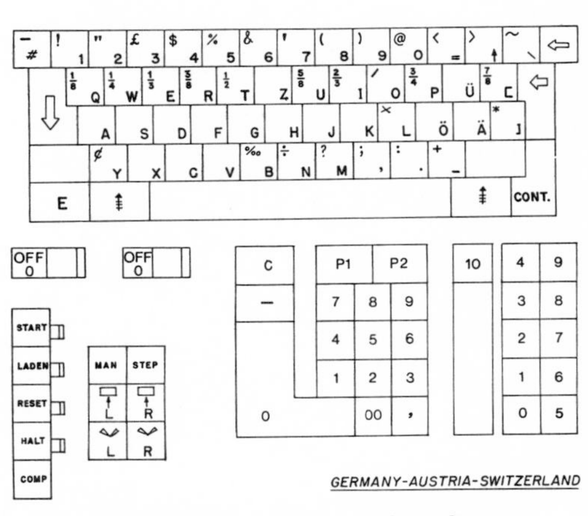

Fig. 6. NCR 399 keyboard for Germany, Austria, and Switzerland.

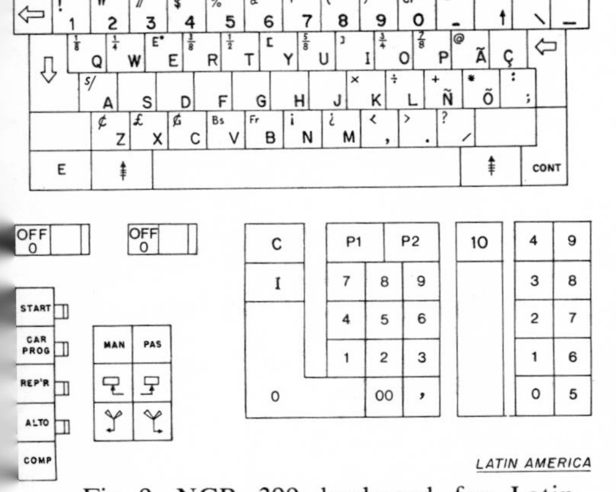

Fig. 9. NCR 399 keyboard for Latin America.

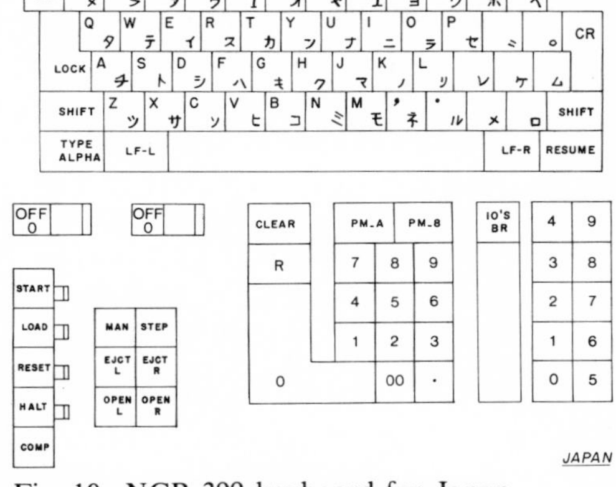

Fig. 10. NCR 399 keyboard for Japan.

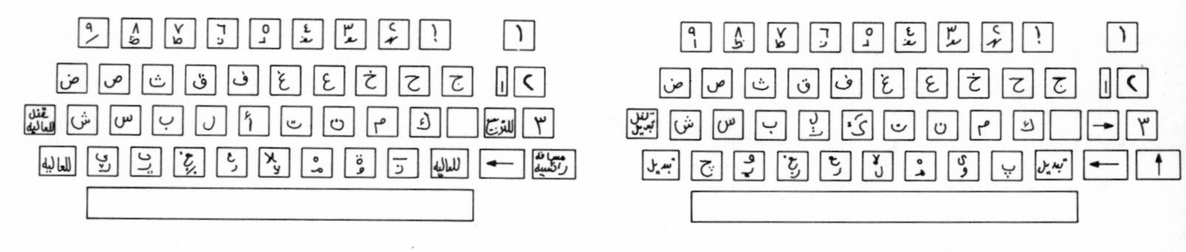

Fig. 11. NCR 395 typewriter keyboard for the Arabic language.

Fig. 12. NCR 395 typewriter keyboard for the Iranian language.

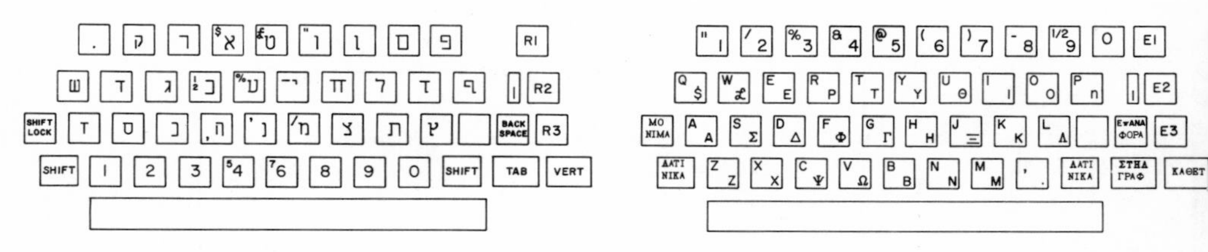

Fig. 13. NCR 395 typewriter keyboard for the Hebrew language.

Fig. 14. NCR 395 typewriter keyboard for the Greek language.

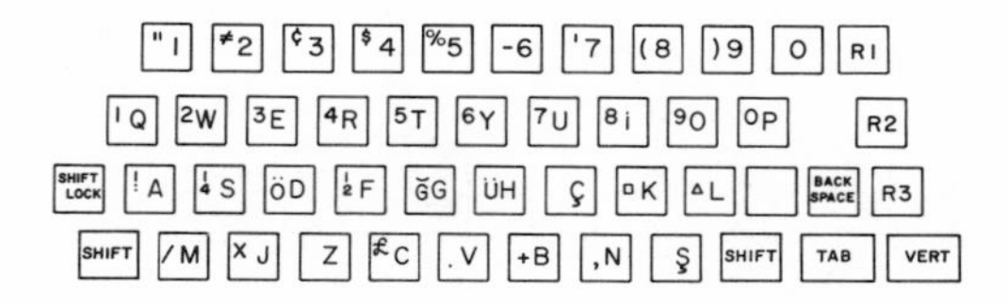

Fig. 15. NCR 33 typewriter keyboard for the Turkish language.

(Fig. 1), but not on the layout for the United Kingdom (Fig. 3). There are also wide variations among countries in symbol locations. The question mark (?) is found in different locations on the South African (Fig. 4) and French (Fig. 5) keyboards.

Graphic symbols are used more extensively in some regions than others. The labeling of the eject and open keys on the Latin American (Fig. 9) and American (Fig. 1) keyboards illustrates the contrast.

The NCR 399 system uses common sizes and arrangements of key tips for all regions except Japan (Fig. 10). On the Japanese keyboard two keys are omitted (upper right corner of typewriter cluster), and the resume and shift keys are shaped differently.

INTERNATIONAL STANDARDIZATION

The ISO, representing more than fifty national standardization bodies, has issued at least three ISO recommendations related to keyboards, covering the layout of typewriter keys (R 1091), the numeric section of ten-key keyboards (R 1092), and certain key-tip symbols (R 1093). These were developed by the ISO Technical Committee on Office Machines (TC95) and approved by at least 60 percent of the national bodies. ISO recommendations are only guidelines. The only valid standard for a country is the national standard of that country.

1. *ISO Recommendation R 1091—Layout of Printing and Function Keys on Typewriters, 1st Edition, June 1969.* This recommendation defines the arrangement, number, spacing, and location of the printing keys and of some of the function keys on typewriters, irrespective of the size of the typewriter. The assignment of numeric and alpha characters to specific keys within the key layout is not part of the recommendation.

2. *ISO Recommendation R 1092—Numeric Section of Ten-Key Keyboards for Adding Machine and Calculating Machines, 1st Edition, June 1969.* This recommendation establishes the composition and layout of ten-key keyboards for adding and calculating machines, as well as the shape of the keys, the slope of the keyboard plane, the maximum keystroke, and the spacing of keys. It applies only to the numerical keys that constitute ten-key keyboards and not to function keys employed with such a keyboard.

3. *ISO Recommendation R 1093—Keytop and Printed or Displayed Symbols for Adding Machines and Calculating Machines, 1st Edition, June 1969.* This recommendation establishes keytop and printed or displayed symbols to be provided on adding machines and calculating machines. It does not prescribe the style of symbols.

A draft ISO proposal entitled, "Guidelines for the Harmonization of General Purpose Alphanumeric Keyboards," is in circulation. This proposal contains guidelines for the layout of general purpose alphanumeric keyboards when they are intended to implement sets of characters based on the Roman alphabet. Keyboard layouts complying with these guidelines will also conform to the requirements of draft ISO Proposal 2126 entitled, "Basic Arrangement for the Alphanumeric Section of Keyboards Operated with Both Hands." Proposal 2126 defines the arrangement of a basic core for the alpha-numeric section of keyboards. It is intended for keyboards to be operated by touch with two hands. In addition to the characters covered by Proposal 2126, the guidelines proposal specifies (1) the pairing and the allocation of a certain number of symbols and punctuation marks, and (2) the allocation to specific keys of additional alphabetic letters (national alphabetic extenders) when they are required, and/or of the separate diacritical or accent marks.

The European Computer Manufacturers Association (ECMA) is also involved with keyboard standardization. This organization published "ECMA Standard for Keyboards Generating the Code Combinations of the Characters of the ECMA 7 Bit Coded Character Set," Standard ECMA-23, in June 1969.[2] This standard defines three keyboard layouts for data processing machines generating the code combinations of the ECMA seven-bit code. One layout, identified as Type A, is a ten-key keyboard with a very limited number of additional keys. It is for use with numeric data. The arrangement may be based on either ISO Recommendation R 1092, which has the adding machine numbering scheme, or the Consultative Committee on International Telegraphy and Telephony Recommendation E.161 for touch tone telephones.[3]

The Type B layout is for use with alphanumeric data that are predominantly numeric. A block of numeric keys (similar to the Type A layout) is located at the right hand end of the keyboard, and the remainder of the keyboard will be as similar as possible to the Type C layout.

The Type C layout is for use with alphanumeric data that are predominantly alphabetic. Its layout will be consistent with those in recommendations by the ISO Technical Committee on Office Machines.

ECMA Standard 23 does not define physical factors, such as key spacing, keyboard slope, size and shape of keytops, nor the way in which the keytops are labeled. ECMA has also proposed a Japanese Kana-alphabet standard for typewriter keyboards implementing 96 and 110 Katakana character sets.

COMMONALITY VERSUS CUSTOMIZATION

There are conflicting views on the advisability of providinƒ a common or customized keyboard arrangement for each unique region. Where language and culture permit, keyboard commonality is advantageous for the user. It simplifies operation, minimizes the need for operator training and retraining on keyboards of different makes or types of machine, and facilitates the exchange of operators within and across national boundaries. The disadvantage of keyboard commonality is that concessions may have been made that reduce job performance and increase operator strain and fatigue due to unnecessary mental processing and unnecessary hand and finger motions.

For economic reasons, manufacturers of business equipment prefer to have a minimum number of specialized keyboards for a given machine. Each additional keyboard configuration costs money for tooling, manufacture, inventory, shipping, and the preparation of specialized training and maintenance manuals. If the potential market is large, the extra costs of customized keyboards can be amortized across many machines with a resultant low additional cost per machine. Customizing the keyboard layout for a country with a small market potential is less likely because the cost per machine can be significant. A critical question is: How much is it

[2]Copies of the ECMA-23 Standard are available from: European Computer Manufacturers Association, rue du Rhône 114, 1204 Genève, Switzerland.

[3]*Editor's note*: See the paper by Archbold in this volume.

worth to customize a keyboard arrangement for a country with limited sales potential?

The approach followed by some manufacturers has been to provide a basic arrangement of keys. The keys may be grouped in clusters, as, for example, the keyboard in Figure 1, based on logical sequences required by the principal applications. Special requirements of nations and regions are considered, and the key tips are marked with the appropriate characters in the language of the region. Sometimes, however, there are restrictions on the flexibility available to locate characters on keys. With some keyboard technologies, each key generates a unique code representing a character. It is not always economically feasible to transform a key code to represent some other character.

Recent technological developments may allow more keyboard customization than in the past. These developments include: (1) transparent key caps that permit interchangeable nomenclature; (2) removable key tips; (3) programmable machine logic that permits a key to change value through simple program and key tip changes; (4) key assemblies in which each key generates a unique, easily changeable code, permitting placement independent of key clusters; (5) key overlays that change the representation of keys according to the overlay; (6) function display keyboards where the key labels automatically change as transactions change; and (7) low-cost keyboards. These developments, however, reduce only some of the costs associated with customized keyboards. Other costs listed earlier must still be considered in any decision involving customization.

Still another consideration in this issue of customization is the number of national and international standards available or under development. As these recommendations become accepted, they must be considered in decisions about customization.

To sum up, the answer to the commonality versus customized approaches to keyboard layout must be sought on a case-by-case basis. In some companies, the solution is developed in trade-off studies. Human factors engineers contribute estimates of fatigue, strain, and performance effects, and customer acceptance; engineers provide technical feasibility and cost information; standards representatives provide information on international and national standards; and marketing representatives provide inputs about customer wants, competition, market potential, and the sensitivity of the market to the additional equipment costs that would be associated with a customized layout. Reaching a final decision about customization requires a judicious balancing of these several factors.

INTERNATIONAL GUIDELINES

The guidelines presented earlier for national keyboard arrangements have been applied to the layout of keyboards for a number of individual countries. The results appear to be satisfactory, although no formal evaluations have compared this approach with other possible ones. Thus, very limited evidence suggests that arrangement guidelines developed for and effective in one nation or culture could be applied successfully in other nations and cultures. Problems may develop, however, when a keyboard must be operated in several regions because of the diverse requirements that may apply to those regions.

In addition to the national guidelines, at least four major guidelines apply to the arrangement of keyboards for international operation.

1. *Optimize the arrangement for each region.* Follow the guidelines presented in the section on national keyboard arrangement in arranging the keys for each region and application. Determine the best possible arrangement(s) for each nation or region independent of the other regions. If there is more than one best arrangement for a region, based on the use of a machine for several different applications, compare the arrangements to identify commonalities and important differences.

2. *Determine the minimum number of key-tip arrangements.* At this point, ignore key-tip legends and determine the minimum number of key-tip layouts that will satisfy the requirements of the various regions. Some important considerations are:

a. The area of the keyboard operated by touch is usually the most important from the standpoint of fatigue and performance.

b. If there is to be a typewriter section on the keyboard, design it to accommodate the maximum number of characters required by any of the countries. That maximum will usually be for a country with national alphabetic extenders, diacritical marks, and national symbols in addition to the basic alphabet.

c. For countries not requiring the extenders, diacritical marks, and special national symbols, the typewriter keys containing such symbols should be assigned to other symbols, controls, and functions having high frequencies of occurrence.

d. Follow international standards, if applicable, unless there is good justification for a different arrangement.

e. Arrange the remaining symbols, functions, and controls to follow the various national arrangements as closely as possible. Where discrepancies exist among countries, provide unique locations by country, if these discrepancies are important enough, or develop the best possible compromise. The compromise may be based on analyses of frequency of use (for example, one country may account for 80 percent of machine use), importance, or logical grouping of functions for all countries using the basic keyboard.

3. *Decide if customization is necessary.* Determine for each region whether fatigue, strain, or performance considerations are sufficient to justify a customized keyboard. The justification for a unique keyboard must be couched in language understandable by groups working toward maxi-

mum commonality. Of course, constraints imposed by keyboard technology may be a factor in establishing the number of unique arrangements possible. If the cost of flexibility is low, then more unique keyboards for a given machine may be possible. Note, however, that minimum retraining and capability for transfer of operators among countries are important user requirements. Keyboard customization for a region, even though economically feasible, may not be desirable from the user's point of view.

4. *Provide key-tip captions in the language of the operator.*

CONCLUSIONS

National guidelines for keyboard design have been applied in individual countries with reasonable success. However, keyboards for individual countries sometimes have variations in language or culture and in historical precedents that tend to interfere with international commonality. The successful development of international keyboards requires that these additional complexities be considered and any conflicts be resolved. International design guidelines have been proposed to aid in reaching a decision on keyboard customization or commonality.

A customized arrangement may be justified for a country if a common layout causes significant fatigue, strain, or performance decrement. However, there is a demand by some users to have an essentially common international keyboard arrangement for a class of machines when operator mobility exists within and among countries, and minimum retraining is important.

Manufacturers generally prefer to minimize the number of different keyboards for a given machine. Additional configurations increase costs that must be passed on to the customer. However, technological developments may make future keyboard customization less expensive than in the past. Even so, user demands for commonality to reduce transfer and retraining costs may result in keyboard commonality when the keyboard cost of customization would be economically feasible. The move toward commonality is bolstered by the interest in and development of international keyboard standards.

Users and manufacturers are becoming more aware of the contributions human factors engineers can make to the arrangement of international keyboards. It is likely that the human factors profession will be invited to participate in such activities even more in the future.

REFERENCES

Alden, D. G., Daniels, R. W., and Kanarick, A. F. Human factors principles for keyboard design and operation—a summary review. Systems and Research Division Report 12180-FR1a. St. Paul, Minnesota: Honeywell, March 1970.

Ancona, J. P., Garland, S. M., and Tropsa, J. J. At last: standards for keyboards. *Datamation*, 1971, **17**(5), 32–36.

Bertelson, P., and De Cae, C. Comparaison expérimentale de deux types de claviers numériques. *Bulletin du C.E.R.P.*, 1961, **10**, 131–44.

Bowen, H. M., and Guinness, G. V. Preliminary experiments on keyboard design for semiautomatic mail sorting. *Journal of Applied Psychology*, 1965, **49**, 194–98.

Conrad, R. Short-term memory factor in the design of data-entry keyboards, *Journal of Applied Psychology*, 1966, **50**, 353–56.

Conrad, R., and Hull, A. J. The preferred layout for numeral data-entry keysets. *Ergonomics*, 1968, **11**, 165–73.

Conrad, R., and Longman, D. J. A. Standard typewriter versus chord keyboard—an experimental comparison. *Ergonomics*, 1965, **8**, 77–88.

Deininger, R. L. Human factors engineering studies of the design and use of push-button telephone sets. *Bell System Technical Journal*, 1960, **39**, 995–1012.

Dvorak, A., Merrick, N. L., Dealey, W. L., and Ford, G. C. *Typewriting behavior.* New York: American Book Co., 1936.

Hanes, L. F., and Baker, G. J. Comparison of ten-key adding machine and push-button telephone numbering arrangements. Operations Evaluation Report Op 7-24. Dayton, Ohio: The National Cash Register Co., December 1965.

Hirsch, R. S. Effects of standard versus alphabetical keyboard formats on typing performance. *Journal of Applied Psychology*, 1970, **54**, 484–90.

Klemmer, E. T. Keyboard entry. *Applied Ergonomics*, 1971, **2**, 2–6.

Kroemer, K. H. E. Human engineering the keyboard. *Human Factors*, 1972, **14**, 51–63.

Mettler, A. J. *Canadian metrication experience*. C.M.A. Publication No. 18. Fonthill, Ontario, Canada: Canadian Metric Association, December 1971.

Michaels, S. E. QWERTY versus alphabetic keyboards as a function of typing skill. *Human Factors*, 1971, **13**, 419–26.

Seibel, R. Data entry devices and procedures. In H. P. Van Cott and R. G. Kincade (Eds.) *Human engineering guide to equipment design* (rev. ed.), Washington, D.C.: U.S. Government Printing Office, 1972.

Strong, E. P. *A comparative experiment in simplified keyboard retraining and standard keyboard supplementary training*. Washington, D.C.: General Services Administration, 1956.

Whitfield, D. British standards and ergonomics. *Applied Ergonomics*, 1971, **2**, 236–42.

United States Department of Defense. *Military standard—keyboard arrangements*. MIL-STD-1280. Washington, D.C.: U.S. Government Printing Office, January 1969.

THIRTEEN

Human Factors Problems in the Design and Evaluation of Key-entry Devices for the Japanese Language

CHARLES R. BROWN

Human factors studies on the design of keyboards are older than the field of human factors engineering itself. Most people generally associate the beginnings of the field of human factors engineering with the 1940s, a period that is some considerable time after pioneering studies of typewriter keyboards by Klockenberg in the 1920s (Kroemer 1972) and by Dvorak and his associates in the 1930s (Dvorak, Merrick, Dealey, & Ford 1936). Following these initial investigations, studies of keyboards, keysets, and data-entry devices have appeared in great numbers. Indeed, Seibel's review of the field (1972) contains no less than eighty relevant citations. Keyboards are the interface between man and a variety of very common machine devices in our society: telephones, typewriters, business machines, and computers of all kinds. Almost without exception, all of the research on the design of keyboards has been done with English-speaking persons using the English language.

Late in 1970, IBM Japan tested two Japanese language key-entry devices using procedures that I developed at IBM's Mohansic Laboratory in Yorktown Heights, New York. These tests brought us face to face with some human factors problems of much greater complexity than would be encountered in comparable work in any European country. At the same time, the complexity of the problem led me to a design principle that I think may have some general applicability.

Charles R. Brown, International Business Machines Corporation, Advanced Systems Development Division, 2651 Strang Blvd., Yorktown Heights, N.Y. 10598, U.S.A.

THE PROBLEM

By far the biggest problem in designing a Japanese keyboard is that it has to accommodate the Japanese language. The basic written language of Japan, called Kanji, was adopted from the Chinese approximately 1700 years ago. It consists of a large, indeterminate number of pictographic characters. For most purposes, however, 2,000 to 4,000 Kanji characters generally suffice. In addition to the Kanji characters, several other classes of character have to be included on a Japanese keyboard. These are:

1. Hiragana, a phonetic subset of about 50 characters used in Japanese syntax
2. Katakana, a phonetic subset of about 50 characters used to transcribe foreign expressions
3. Japanese numerals, the equivalent of 0 through 9 plus a few for large quantities such as 100 and 1,000
4. Arabic numerals, 10 in all
5. The Roman alphabet, consisting of 26 characters
6. Miscellaneous characters consisting of such things as TV channel numbers, baseball symbols, and special numerals for financial transactions.

Imagine trying to accommodate all of these character sets with conventional European typewriter keys. With one character per key, the keyboard would be about 1 meter wide and 1 meter high. Nevertheless, the keyboard can be a small fraction of this size without degrading the legibility of the character legends. However, approaching this theoretical minimum size requires keying techniques very different from those of the European typewriters. Our test keyboards illustrate this point.

THE TEST KEYBOARDS

One keyboard requires a stylus to actuate a key. The dimensions of the individual keys (or character positions) are somewhat over 4 mm, too small to depress with a finger. However, this size is quite adequate if the key is depressed with a pencil-like stylus. A very different keying technique for approaching minimum keyboard size is used with the second keyboard. On each key is a legend of 12 characters. These characters are arranged in an array of 4 rows and 3 columns. To key a character requires the use of two hands. The right hand depresses the key containing the desired character. The left hand selects the desired character from the 12 on the key by depressing one of 12 shift keys arranged in the same pattern as the legend characters. Because the shift keys are arranged in 3 columns, the little finger of the left hand is not used for shift-key operation.

KEYING TIMES

Naturally, we were greatly interested in the performance, that is, the keying times and error rates, associated with these two keyboards. Would the greater complexity of these devices, in comparison with conventional American and European devices, make a difference in performance? I shall use the concept of a word for purposes of comparison. Two characters, I am told, are roughly equivalent to one word in Japanese. Five characters are usually taken as the average word length in English. Using the card punch as a standard for keying English words, Klemmer and Lockhead (1960, 1962) report performance values for trained operators of:

- Keying times: 0.29 sec/character, or 1.45 sec/word
- Error rates: 0.05% characters in error, or 0.25% errors/word.

Comparable performance values for the Japanese shift keyboard in the application environment are:

- Keying times: 0.75 sec/character, or 1.5 sec/word
- Error rates: 0.2% characters in error, or 0.4% errors/word

Clearly, on a word or informational basis the speeds and error rates in both English and Japanese are similar. From this we conclude that great cultural variation has produced little or no variation in the level of skilled performance.

There are no comparable data on the use of the stylus keyboard in the application environment. However, let us take a look at what was learned about both keyboards in the test environment and how well the test results correlate with the results found in the application environment.

LAYOUT OF THE CHARACTERS

In the tests of the two Japanese key-entry devices, we decided to use the layout of the shift keyboard on both devices. The stylus keyboard had not yet been used in an application environment,[1] whereas the shift keyboard, called the Kantele (Kanji teletype), had been used for over twenty years in the newspaper industry. Fortunately, the Kantele layout could be placed on the stylus keyboard's physical configuration.

Figure 1 shows that the keys on the Kantele keyboard are arranged in quadrants consisting of 4 rows and 12 columns. Each key contains a 4- by 3-character array. The 12 shift keys are located in the lower left-hand corner. In this same general area are also some function keys that were not

[1]The stylus keyboard, an experimental device, was designed and developed by IBM's Advanced Systems Development Division in Yorktown Heights, New York (Juliusburger, Krakinowski, & Stilwell 1970), and was demonstrated at the World's Fair in Osaka, Japan, in 1970.

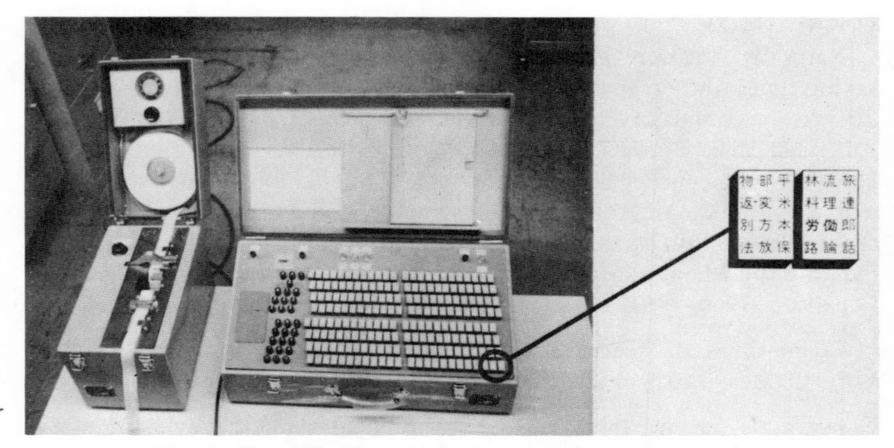

Fig. 1. The shift (Kantele) keyboard and paper tape unit.

used in the present study. Figure 2 shows the stylus keyboard configuration. Our interest is limited to the 2 large modules in the center. Each module has 8 single-character key arrays of 5 rows and 40 columns. A character key of the stylus keyboard is not similar to a conventional, mechanical key. A nomenclature sheet covers the switch positions. A character is keyed by depressing this nomenclature sheet approximately 1 mm with a stylus. Auditory feedback is provided by the cardpunch.

The result of placing the characters of a shift keyboard quadrant on four of the character arrays of the stylus keyboard is illustrated in Figure 3.

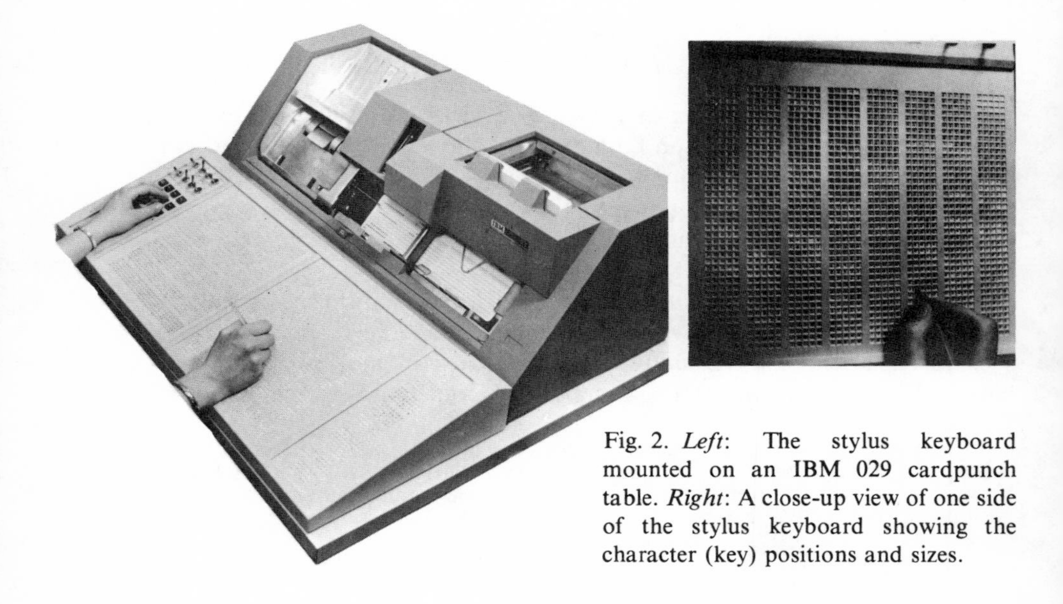

Fig. 2. *Left*: The stylus keyboard mounted on an IBM 029 cardpunch table. *Right*: A close-up view of one side of the stylus keyboard showing the character (key) positions and sizes.

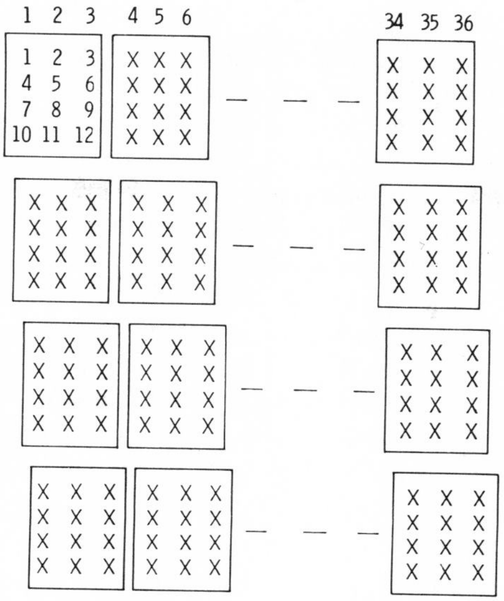

SHIFT KEYBOARD QUADRANT

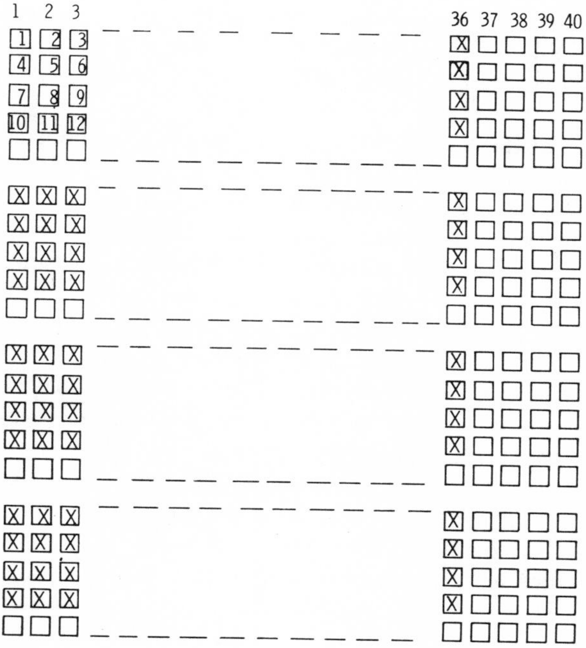

STYLUS KEYBOARD QUADRANT

Fig. 3. The shift keyboard layout placed upon the stylus keyboard configuration—one quadrant.

Placing one Kantele row on one stylus character array maintains the vertical perceptual grouping. Appropriate color coding of the stylus keyboard maintains the horizontal grouping of the separate keys of the Kantele. If the shift keyboard layout is placed on the stylus keyboard as indicated, a number of stylus keyboard key positions are not used. The fifth row of each character array is empty. Also, the right-most four columns of each character array are empty. A Kantele quadrant is 181 mm wide and 95.3 mm high. Placed upon the stylus keyboard, a quadrant increases in size to 217.8 mm wide and 135.8 mm high. The somewhat greater width of the stylus keyboard qudrant is due to the somewhat larger character area. The appreciably greater height of the stylus keyboard is due to the larger character area, the unused character rows, and the separation between character arrays. On both keyboards we are concerned with a total ensemble of 2,304 characters.

OPTIMUM LAYOUT

To get a fair test of these two physically dissimilar devices, we needed to assure ourselves that we had nearly optimum layouts. Performance by highly skilled operators on the Kantele keyboard in the application environment suggests that its layout is nearly optimum. Text is keyed at 1.00 sec/character after a few months and at 0.75 sec/character after 1 year. After 1 year, only 0.2 percent of strokes are in error. As previously noted, this is roughly equivalent to the performance of an operator trained on a conventional card punch in the United States.

An analysis of the Kantele's layout into quadrants supported our feeling that the Kantele layout is close to optimum. The primary organization is by usage. The characters with the highest usage are located in the lower right-hand quadrant. Usage decreases from quadrant to quadrant in a counter-clockwise fashion. The human factors benefit from this organization is that the average area searched and the average distance reached are greatly reduced, so that the effective size of the keyboard is greatly reduced. The high-usage region contains far less than half the characters, but these characters are used far more than half the time. The odds are high that a character picked at random will be in the high-usage region. Figure 4 shows the cumulative usage accounted for by Kanji characters rank-ordered from the most used to the least used. The 200 most used Kanji characters account for over 50 percent of the usage.

To pursue the benefits of organizing a keyboard by usage, consider first a layout in which the entire keyboard (containing all the Kanji characters) is one region. Now rearrange the characters into two separate regions, one containing the characters used more often and the other the characters used less often. Since search and reach in the high-usage region of the two-region keyboard are less than those in the one-region keyboard, the rate of keying within the former will be improved over the keying rate that existed

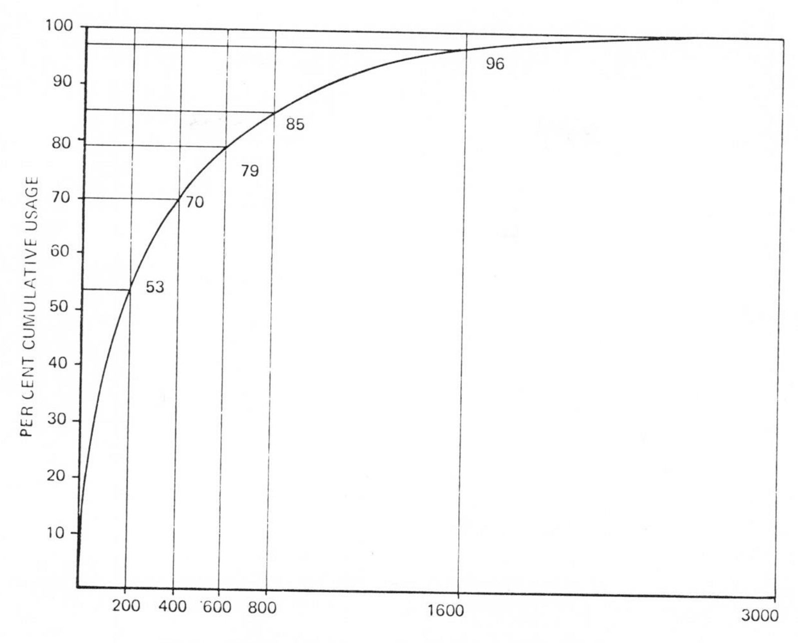

Fig. 4. Percent of usage accounted for by Kanji characters rank-ordered from most used to least used.

before the keyboard was partitioned. This saving in the time to key a character within the high-usage region is amplified by the odds that a character is in that region to produce a gain in time available to key within the low-usage region. This gain results in an improved keying rate for the keyboard as a whole, since no additional time is actually required to key a character within the low-usage region.

SIZE OF THE HIGH-USAGE REGION

How large should the high-usage region be? The answer involves a trade-off between savings and odds. With a small high-usage region, the odds that a character will be in that region may be so negligible that it would attenuate appreciable savings in keying speed. With a large high-usage region, the savings in keying speed may be so negligible as to attenuate appreciable odds. Thus, the layout design problem is to estimate a size of high-usage region that will have *moderate odds together with moderate savings in keying speed.* The Kantele layout appears to be rather close to the optimum solution to this design problem.

Quantitatively, these ideas take the following form. Let the average time to key a character before the keyboard is partitioned into a high-usage region, R_1, and a low-usage region, R_2, be K. Maintaining the overall keying time, we assume that the time to key a character within R_1 is reduced by an amount Δ_1. This results in an excess time, Δ_2, to key a character in R_2. Therefore, the average time (K) to key a character on the entire keyboard is expressed by:

$$P_1 (K - \Delta_1) + (1 - P_1)(K + \Delta_2) = K, \tag{1}$$

where P_1 is the probability of keying in R_1. Solving for the gain, Δ_2, in terms of savings, Δ_1, we have:

$$\Delta_2 = \left(\frac{P_1}{1 - P_1} \right) \Delta_1 . \tag{2}$$

Equation (2) says that the gain is the savings amplified by the odds of keying in the high-usage region. For example, if the savings were 1.0 sec and the odds were 9 to 1 ($P_1 = 0.9$), the gain would be 9.0 sec.

However, since K is sufficient time to key anywhere on the unpartitioned keyboard, it is sufficient time to key a character within R_2, given that an operator has learned whether or not a character is in R_1. Thus, we can calculate the minimum improvement in keying time to be expected over the entire keyboard by setting equation (1) equal to K_n, the new keying time, and Δ_2 equal to 0. This gives us:

$$\frac{K_n}{K} = \frac{K - P_1 \Delta_1}{K} . \tag{3}$$

This expression shows that as the usage in R_1 approaches unity, the average time to key a character for the keyboard as a whole approaches the time to key a character in R_1. Figure 5 illustrates this graphically. The axes may be interpreted as follows:

K_n / K = Either (a) the average time (sec/character) to key a character on the partitioned keyboard relative to the average time to key a character on the unpartitioned keyboard or (b) the keying rate (characters/sec) for the unpartitioned keyboard relative to the keying rate for the partitioned keyboard. As this number decreases from 1.0, keying on the partitioned keyboard improves relative to keying on the unpartitioned keyboard.

$(K - \Delta_1)/K$ = Either (a) the average time (sec/character) to key a character within the high-usage region relative to the average time to key a character for the unpartitioned keyboard or (b) the keying rate (characters/sec) for the unpartitioned keyboard relative to the keying rate within the high-usage region. As this number decreases from 1.0, keying in the high-usage region improves relative to keying on the unpartitioned keyboard.

Recall that to achieve a $P_1 = 0.9$ or so requires relatively few of the characters to be in R_1. Approximately 800 Kanji and 50 Hiragana characters account for 90 percent of the usage for standard text. The keying time in a high-usage region consisting of these characters would essentially determine the average keying time for a keyboard in excess of 3,000 characters. So the high-usage region need not exceed approximately 25 percent of the keyboard. But how small can the high-usage region be? Note that if $P_1 = 0.5$, equation (3) shows that the keying within R_1 must take no time at all if the over-all average is to be 0.5 of K (that is, $\Delta_1 = K$). Thus, R_1 must contain a large enough set of high-usage characters so that the prob-

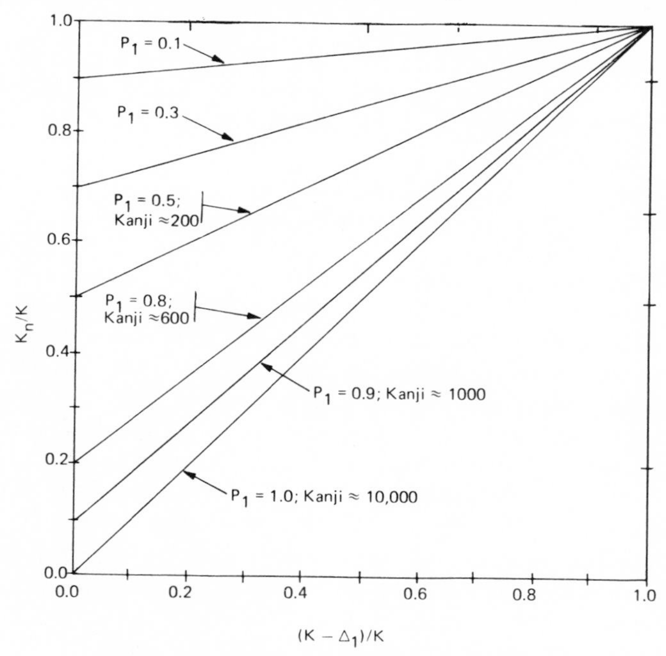

Fig. 5. Improvement of keying throughout the total keyboard as a function of the improvement of keying in the high-usage region.

ability of keying in the area is substantially greater than 0.5. The Kanji usage curve (Fig. 4) in conjunction with Figure 5 locates a relatively narrow range within which the optimum lies. My judgment is that an R_1 within the range from approximately 600 Kanji characters ($P_1 \cong 0.8$) to 1,000 Kanji characters ($P_1 \cong 0.9$) is nearly optimum.

CHARACTER-SEQUENCE RULE

Also important is the organization of characters within a keyboard region. On the keyboards studied, the Kanji characters within a region are arranged by a pronunciation sequence, the Hiragana. The first syllable of each character's name is used as an index of arrangement. The secondary rule of arrangement is the number of strokes that make up the character. Thus, a dozen or more Kanji having the same first syllable may be grouped together.

The Kanji characters are sequenced in columns. A column is 3 characters wide (one Kantele key) and 16 characters high (four Kantele keys). The order of characters within a column is from left to right across the 3 characters in a line and from top to bottom through the lines of 3 characters. Succeeding columns are to the right. On the stylus keyboard, adjacent columns were of different colors as a perceptual aid. Exceptions to this scan rule occur when other character subsets are within a quadrant. For example, the Hiragana subset is located centrally within the high-usage quadrant. This permits 80 percent to 90 percent of standard text to be keyed from this quadrant.

EVALUATION PROCEDURES

For reasons of time and cost, the evaluation had to be conducted with one keyboard of each type. To approach the level of training of full-time operators in the application environment, each operator in the evaluation was trained on a mock-up and tested on the associated keyboard. The mock-ups were essentially just layouts of the character positions. The similarity between mock-up and keyboard was high for the stylus keyboard, since key depression on the keyboard is minimal. The similarity between the mock-up and the shift keyboard was low, except for character position. Shift-key training was limited to locating the shift-key position by touch.

Training sessions were of two types: general practice, in which regions of the keyboard were studied for periods of time proportional to the average usage of characters in the region, and document practice, in which four 40-character documents were rehearsed for 50 minutes. Both types of practice sessions occurred in the morning and afternoon.

Testing also occurred both in the morning and afternoon. The four documents rehearsed in the preceding practice session were timed. Two of these documents were segments of text, a mixture of pictographic Kanji and phonetic Hiragana. One document was text with Kanji deleted—pure Hiragana. One document was text with Hiragana deleted—pure Kanji. A similar set of four unrehearsed documents was timed during test sessions. These new documents became the documents for the next practice session. Unrehearsed text was not tested until the second week, when layout familiarity permitted its inclusion within the scheduled time period. The operators had immediate knowledge of their time scores. Information on their error scores was delayed by two or three days.

Ten weeks of evaluation were planned. However, one week of data collection was lost in general orientation procedures. Five girls were hired for each keyboard. We lost one operator from each group: one resigned, and the other was absent so much that her data could not be used. The groups were matched according to a standard card-punch operator examination.

RESULTS

UNREHEARSED TEXT

For keying times no statistically significant difference was obtained between the learning curves for the two keyboards. At week 2, keying was somewhat under 3.50 sec/character; at week 9, the keying time was somewhat under 2.00 sec/character. When the test ended, the improvement curve for keying time had still not approached its asymptotic value. Yet the improvement in keying time was substantial, and the level finally reached appears to be fairly close to what probably occurs in the application environment.

The shift keyboard's error rate for unpracticed text was substantially higher than the asymptotic value of the application environment. This is reasonable. Since insufficient shift-key training preceded this evaluation, the test procedures did not simulate mastery. Throughout the test, the stylus keyboard's error rate for unpracticed text remained near the asymptotic value characteristic of the application environment.

PRACTICED TEXT

Keying time. Figure 6 shows the improvement in keying time for both keyboards for Hiragana, text, and Kanji. The most striking feature of these data is the rapid improvement during the first four weeks and the ensuing stability, as shown by the gradual improvement, after week 4. For the Hiragana and Kanji curves, this means relative stability after approximately eight hours of document practice. For the text curves, this means relative stability after approximately sixteen hours of document practice.

Another striking feature of these data is their regularity. First, the Hiragana, text, and Kanji curves for a particular keyboard never cross. For every week, the data are ordered by size of the character set: Hiragana fastest, then text, and Kanji slowest. Second, corresponding curves for the two keyboards never cross. The stylus curve is *always* below its corresponding shift-keyboard curve. The third striking feature of these data is the rapid approach of the stylus keyboard's curve for text to the asymptote characteristic of the application environment. At week 4, keying with the stylus keyboard is at 0.99 sec/character. At week 9, keying is at 0.87 sec/character.

Last, but not least, note the fairly small increase in keying time with the increase in size of the character set. After week 4, for example, text is

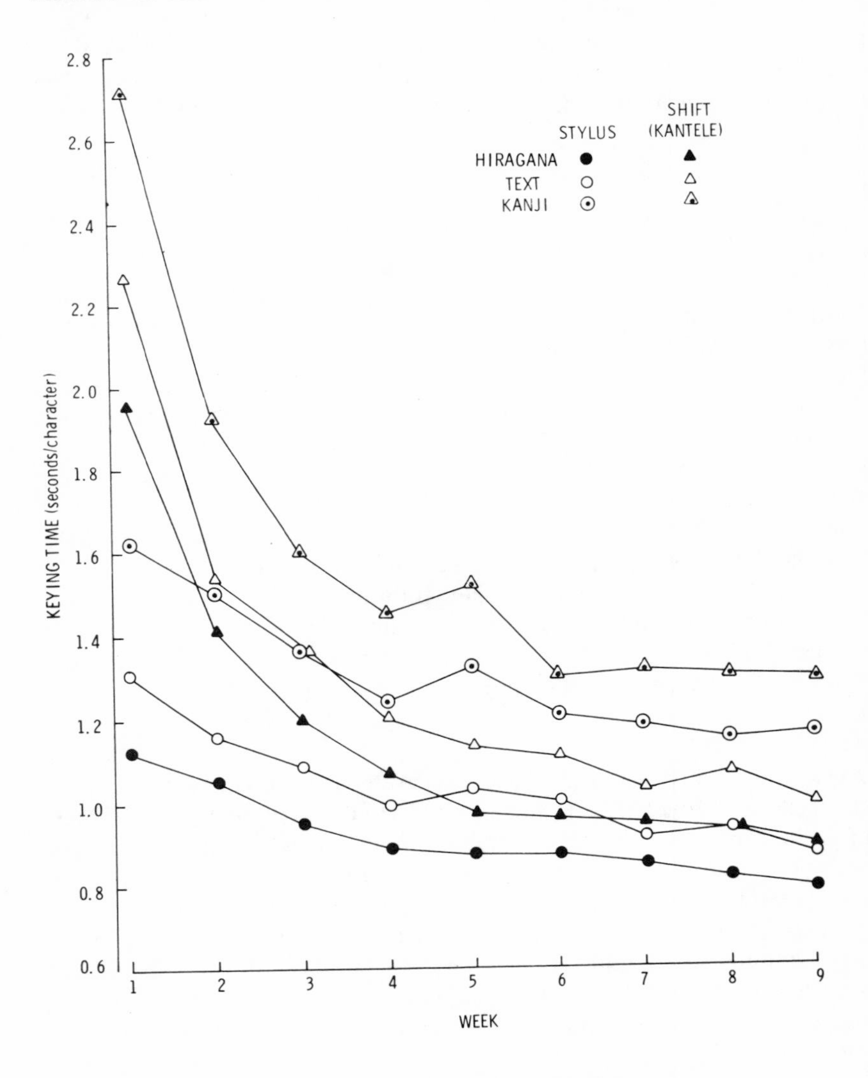

Fig. 6. Keying times for practiced material.

keyed about 0.10 sec/character slower than Hiragana. Figure 7 shows the keying time for text as a linear function of the keying time for Hiragana. Table 1 gives the results of the correlation and regression analyses on those keying times. The same linear function fits the data for both keyboards. Indeed, the fit is to within 1 percent of the total variance. The function through the Kanji keying times is drawn parallel to the curve through the text keying times to show the similarity between the two sets of data.

Error rate. Shift-key operation was not mastered during the course of the evaluation. Figure 8 shows erratic improvement for the shift keyboard

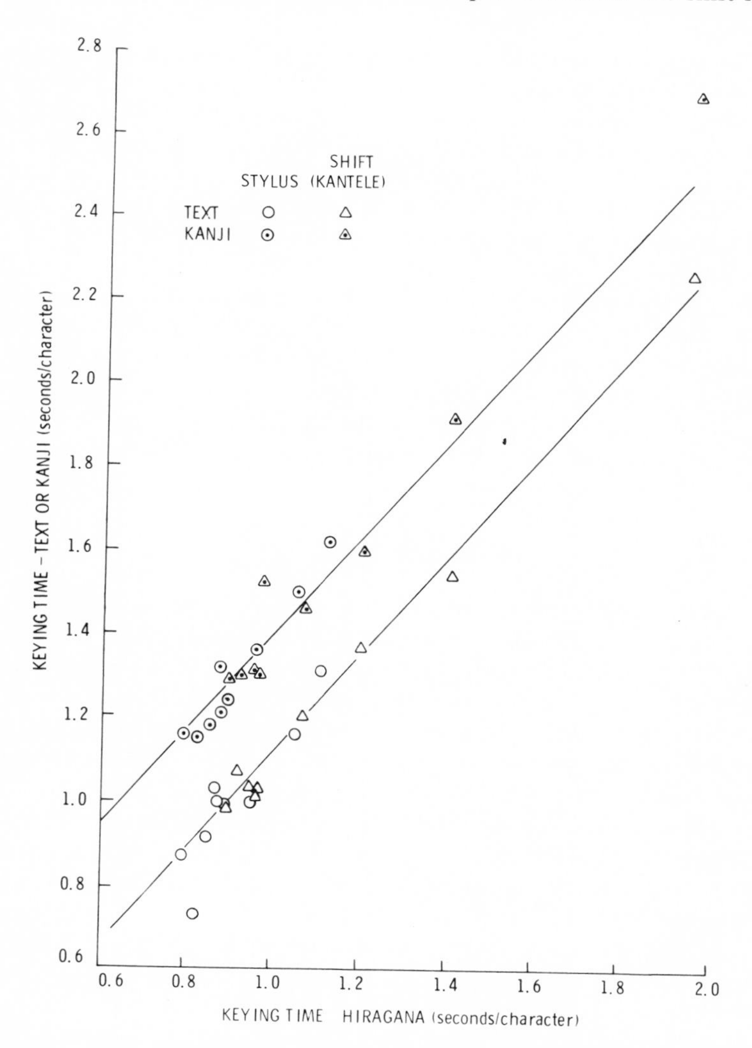

Fig. 7. Keying times for text and pure Kanji as a function of the keying time for pure Hiragana.

Table 1. Correlation and regression analysis for the data in Figure 7. The data for both keyboards have been combined

Variable			
X	Y	r^2	Regression equation
Hiragana	Text	0.9896	$Y = 1.16X - 0.03$
Hiragana	Kanji	0.9779	$Y = 1.32X + 0.09$

throughout these tests, and the final level is not near the asymptote usually found in the application environment. This is reasonable because the shift-key training that preceded the tests was lacking in realism, and the testing itself did not simulate mastery.

By contrast, the median error rates for the stylus keyboard were near the asymptote typical of the application environment (see Table 2). This is reasonable, since the stylus keyboard does not require learning the shift-key operation. Because the limiting factor in performance is the size of the character set and not the shift-key operation, in a sense the stylus keyboard simulates shift-key mastery.

A finding of some interest is that the error rate did not increase with the size of the character set for practiced material. Figure 9 shows text error predicted from Hiragana error. Table 3 shows that for both keyboards combined, errors in Hiragana can predict well over half (58 percent) of the variance in errors in text. The identity function ($Y = X$), is drawn to indicate that error rate is almost entirely insensitive to the size of the character set.

DISCUSSION

If I had not observed professional operators perform on the 12-shift keyboard, I probably would have overestimated the time required to key a character. The speed with which work can be done on such a keyboard is almost unbelievable. The operators are impervious to distractions and noise and are able to read, seemingly without difficulty, hand-written script with marks of correction. Skilled performance of such a complex task is truly impressive.

The excellent keying rate on the 12-shift keyboard is certainly not accidental. Mastering 12 shifts takes time, but it can be done. By exploiting the adaptability of the human, the size of what would otherwise have to be a very large keyboard has been greatly reduced. Over time, learning compensates for the complexity of the shift-key keyboard. However, the very large physical size of the keyboard is a permanent, limiting factor on performance.

The stylus keyboard is a solution to the problem of keyboard size that is quite different from the multiple-shift approach. It presents no insoluble

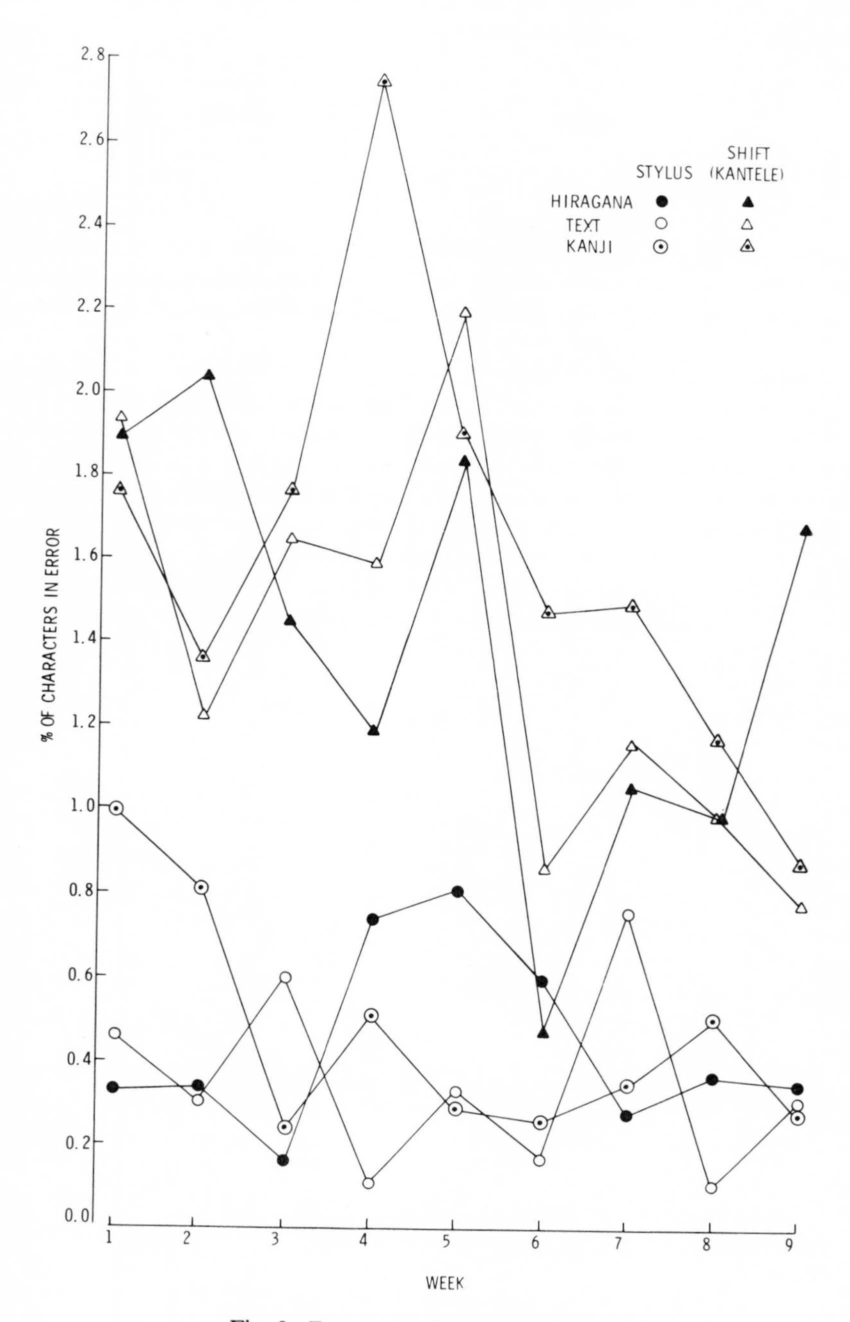

Fig. 8. Error rates for practiced material.

Table 2. Measures of central tendency of the percentages of characters in error for the stylus keyboard (see Fig. 8)

Test material	Mean	Approximate median
Hiragana	0.43	0.36
Text	0.36	0.25
Kanji (last 7 weeks)	0.26	0.25

problems and essentially eliminates the error-learning curve characteristic of shift-key operations. Furthermore, stylus operation permits practiced material to be keyed somewhat faster than does shift operation.

KEYBOARD ORGANIZATION

In addition to comparing the shift and stylus operations necessary to reduce the size of the keyboard, we have made use of a principle of keyboard organization by usage. The exploitation of this statistical property of the language markedly enhances the keying rate. Comparable examples of reducing the complexity of difficult tasks would be welcome additions to our reference works on human factors engineering.

Even though the layout of the keyboards into usage regions reduces the average distance between successive characters of text, we are still sur-

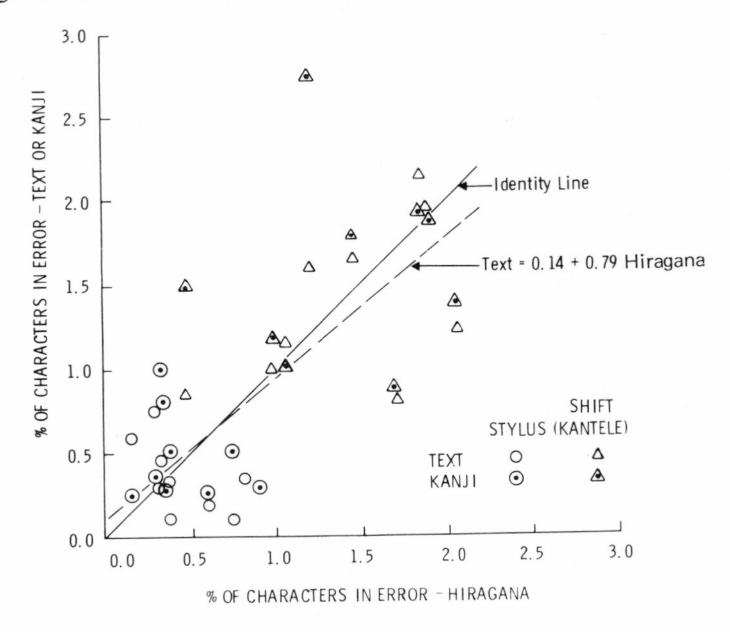

Fig. 9. Error rates for text and pure Kanji as a function of the error rate for pure Hiragana.

Table 3. Correlation and regression analysis for the data in Figure 9. The data for both keyboards have been combined

Variable			
X	Y	r^2	Regression equation
Hiragana	Text	0.583	$Y = 0.79X + 0.14$
Hiragana	Kanji	0.402	$Y = 0.73X + 0.37$
Hiragana	Text and Kanji	0.473	$Y = 0.76X + 0.26$

prised by the relative insensitivity of performance to the size of the character set once the layout by usage has been established. Hiragana is keyed about 10 percent faster than text, whereas Kanji is keyed about 20 percent slower than text. Clearly, our layout must be nearly optimum, for no layout would be expected to reduce the keying of text below that of a small 50-character subset, that is, the Hiragana. Thus, the size of the character-set, which determines the duration of learning, may have little effect upon the final level of skilled performance. It appears that the difficulty of a task or device is more appropriately measured by the time it takes to acquire the skill to perform the task, or operate the device, than by the level of asymptotic performance achieved on the task or device.

SHORTCUTTING TRAINING

Our attempt to simulate the performance of a skilled operator was successful. Our operators, after only a relatively few hours of document practice, exhibited the kind of stable keying performance one expects of truly skilled operators. The stylus keyboard operators approached very nearly the keying rate of operators in the application environment. This occurred even though our operators (a) were temporary employees rather than professionals, (b) were of a different sex from operators who normally do this kind of work, and (c) spent most of the day working with a mock-up that did nothing. Given the many differences between the test and application environments, our estimate of skilled performance is remarkably good.

The rapid improvement and ensuing stability of our data are encouraging. If we can show that other complex skills can be estimated with techniques similar to those used here, we will be able to evaluate the influence of various experimental conditions upon skills in 1 percent or so of the normal time required to acquire the skill.

SUMMARY

Japanese language keyboards may contain 100 times as many characters as European typewriters. If the keyboard is to approach the minimum size consistent with legibility requirements for the characters, keying must involve stylus operation or multiple shifts, or both. The average area searched and the average distance reached can be reduced further by grouping the characters into two or more regions by usage. In that way, the odds can be increased that a character of text will be found in a relatively small, high-usage region. That in turn reduces the effective size of the keyboard. The use of either keying technique in conjunction with the layout principle results in performance comparable to that typical of skilled American card-punch operators. The enormous difference between the character sets and the keying techniques in the two cultures changes neither the average number of words keyed per unit of time nor the average number of errors made per word.

To estimate the performance level of fully trained operators, practiced material was used to simulate memory for character position. A few hours of such practice resulted in a level of performance that normally takes months to acquire. Thus, the asymptotic level of keying the Japanese language may be estimated without actually training operators to that level.

REFERENCES

Dvorak, A., Merrick, N. I., Dealey, W. L., and Ford, G. C. *Typewriting behavior*. New York: American Book Company, 1936.

Juliusburger, H. Y., Krakinowski, M., and Stilwell, G. R., Jr. The Kanji entry feature. Technical Report No. 17-245, IBM Advanced Systems Development Division, Yorktown Heights, New York, March 1970.

Klemmer, E. T., and Lockhead, G. R. An analysis of productivity and errors on keypunches and bank proof machines. Research Report RC-354, IBM Research Center, Yorktown Heights, New York, November 1960.

Klemmer, E. T., and Lockhead, G. R. Further data on card punch operator performance. Research Note NC 39, IBM Research Center, Yorktown Heights, New York, 1962.

Klemmer, E. T., and Lockhead, G. R. Productivity and errors in two keying tasks: A field study. *Journal of Applied Psychology*, 1962, **46**, 401–8.

Kroemer, K. H. E. Human engineering the keyboard. *Human Factors*, 1972, **14**, 51–63.

Seibel, R. Data entry. Chapter 7 in Van Cott, H. P., and Kincade, R. G. (Eds.) *Human engineering guide to equipment design* (rev. edition), Washington, D.C.: U.S. Government Printing Office; 1972.

FOURTEEN

Some Problems of Voice Communication for International Aviation

H. P. RUFFELL SMITH

Comprehensive and rapid communications are indispensable to the effective operation of modern aircraft. In the days when transport flights were few and slow it was possible to use existing public telegraph systems or private cable companies for communication purposes. This method did not impose serious language problems, because many messages were stylized and codes could be produced in any language. In addition, messages were received by an operator who had no other work and who was able to record them in writing. If there was any doubt as to the meaning of what he had received, an operator could refer the written word to another person without the intervention of errors from the malfunctioning of short-term memory.

As air traffic increased in amount and speed, telegraph systems were no longer adequate to its needs. At the same time, improvements in the technology of long distance voice transmission saw the gradual replacement of telegraphed messages by telephone conversations. Telephones have a number of advantages: they are flexible, easy to use and have cheap terminals. Offsetting these advantages are several important disadvantages. The transmission of speech over telephone lines is a time-consuming way of transmitting data that can often be done more quickly by data transmission systems. Telephone circuits, whether they be composed of cables or microwave links, involve high capital cost. Such expensive systems can transmit much more information if messages are efficiently encoded in ways that do not involve speech. Telephone conversations are liable to

H. P. Ruffell Smith, 55 Hallam Court, London W.1, England.

error and, although they can be recorded for future reference, no practical system has yet been devised that will enable an operator to locate quickly a particular conversation when it is needed for recall. Finally, when telephonic communication is used internationally, problems of language impose an additional selection and educational load on a wide variety of airline and government employees.

The disadvantages of telephonic communication have forced airlines to build up high-speed digital communication systems that can interface with the memory and calculating functions of central computers. At the same time a vast world system of telex communications has been evolved and, even when nonstandard situations do not allow the use of stylized inputs and outputs, hard copies of messages are available without the need for a telegraphist.

The situation is, however, different with regard to ground-to-air, air-to-ground, and air-to-air links. In the early days, voice communication for these purposes was unsatisfactory and reliance had to be placed on wireless telegraphy (W/T). Because this manual method of transmitting was so slow, a means had to be found to increase the message content of a small number of characters. The "Q" code was developed in response to that need. It is still sometimes used today, even for voice communication. This code consists of two letters preceded by the letter Q and covers most of the values that need to be transmitted between aircraft and ground stations. One great advantage of the Q code is that it is truly international. Because all nations use Morse code letters, the meaning of any message can be found in any language without recourse to verbal translation.

The main difficulty with W/T systems is that they require a continuous manual input through a key during transmission and enough mental concentration during reception to make it impossible for the operator to engage in any other cockpit duties, especially during the critical phases of aircraft letdown and approach. Improvements in the range of very high frequency (VHF) by the use of forward links, improvements in automatic pilots, and the reduced monitoring requirements of turbine power enabled the communications load to be accepted by other crew members. The airlines were quick to realize this and took advantage of the situation to remove the costly communications operator from the cockpit.

Meanwhile the 1950s saw a vast upsurge in commercial air traffic. To avoid collision risks it became necessary for pilots to report their present position and to estimate the time they would be at their next turning point. This was made possible by the ability of pilots to talk directly to air traffic controllers by VHF radio. Thus communications became more than a simple information service, because through them aircraft were controlled from the ground in situations where there were other aircraft or where there was high ground in proximity. Under these circumstances, mistakes in communication or interruptions to it could have dangerous results.

THE LANGUAGE PROBLEM

At the present time, the control of air traffic requires a common communication system capable of providing the crew of an aircraft in flight with the information it needs. Currently this need is met by a voice system. Despite serious shortcomings this system has persisted for more than two decades, probably because of the relative cheapness of its ground and air terminals. The different languages of the many nations engaged in international traffic are an obvious impediment to the use of such a system. This impediment has been recognized in the Western world and has resulted in agreement to use a common language for all aircraft operating purposes. English was chosen because in the immediate postwar years the majority of aircraft involved in international flights were operated by English-speaking countries. The pilots of many European airlines at this time were either British or American or had been trained by instructors whose native tongue was English.

Thus in theory the stroke of an administrator's pen solved the communication problem between pilots and air traffic controllers of all the Western nations. Unfortunately there are wide variations in the pronunciation and usage of the English language. Even within the confines of the British Isles some local accents, as, for instance, the dialect in the Newcastle area, are not understood by the majority of the population.

However, within the British Isles there is a reasonably standard pronunciation that has been used by most radio announcers and is universally understood by anyone who has the necessary education to become a member of air crew or air traffic controller. This situation does not apply internationally, because English spoken by other nations may vary widely. Marked differences in pronunciation are apparent both when English is the native language of the speaker, as in the United States, or where it is spoken by people who have another native language. Such individuals may consider themselves word perfect in English, but their pronunciation may be such that members of other nations may have extreme difficulty in comprehending them.

Even if mistakes in meaning do not occur, communication between an Englishman and others speaking the language in its various forms and accents may be slowed down. However, when English is used as a common language between either those for whom the language is not native or when the native accent of one is extreme, the meaning itself may be lost or misunderstood. In civil aviation most of the difficulties have been overcome by the use of stylized messages with which both air and ground personnel become familiar. Nonetheless, the basic problem remains and becomes manifest during abnormal circumstances. Unusual situations are often dangerous. In such cases accurate and rapid communication can be a prime necessity.

Evidence that communication is implicated in aircraft safety is hard to come by. A few years ago, however, I examined the human factors surrounding a group of accidents in which aircraft collided with high ground (Ruffell Smith 1968). This kind of accident is responsible for the death of more passengers in civil aircraft than accidents due to any other factor. My analysis seemed to indicate that although difficulties in communication could not be pinpointed as definite causes, they were commonly associated with accidents of this kind.

During a later series of investigations into the workload and working conditions of pilots in civil airlines, I noted the number and kind of errors that occurred from the use of speech for air-to-air, air-to-ground, and ground-to-air communications. The observed errors were more numerous in fast jets, 13 in 238 segments, compared with 3 in 163 segments for slower aircraft. Of these 16 errors, 7 were mistakes in call-signs, 4 mistakes in position-reporting for air traffic control, and 3 errors of interpretation. In every case at least one of the communicators was British. It seems likely that errors would be more frequent when both the speaker and listener did not have English as their native tongue.

Errors of the kind noticed are seldom very important because the mistake is soon discovered and rectified. However, in a small number of cases results could be disastrous as, for instance, when one aircraft responds to the instructions intended for another, due to a mistake in a call-sign. Most of the communication errors associated with call-signs are made by air crew and are due to the similarity between some tail numbers or lettered call-signs of aircraft which happen to be in the same controlled air space. Air traffic controllers do not seem as likely to make this error, because the transmission by a pilot is coded by the individual characteristics of his voice as well as the aircraft call-sign itself.

Although delays in communication may not be as dangerous as errors, they do have some safety implications. In a crowded speech channel every repetition makes the situation worse. The need to take immediate advantage of any gap in the conversation tends to distract the pilot from other cockpit tasks that may be critical, and an air traffic controller is sometimes prevented from giving an urgent instruction because the channel is occupied. For some reason people seem to become garrulous when they are using a strange language.

The pilot also has a continuing need for accurate and up-to-date information about the weather. Nowadays changes in en route weather are of less importance than formerly, since the wing heating equipment of civil transport aircraft can cope with even the worst icing, and wind changes will be immediately sensed by sophisticated navigation equipment. However, the weather at the destination airport or at its alternate is still of vital importance with regard to visibility, wind speed and direction, cloud base, and temperature. All of these weather conditions may be critical in the decision to land or divert, and in strong cross winds the wind vector may be required right up until the aircraft is committed to land.

The requirement for meteorological information is currently met by continuous broadcasts from tapes that are changed at frequent intervals, each geographic area having a separate frequency. There are, however, usually ten to twenty airfields in the area served by one frequency. This means that to obtain the landing weather for his destination or alternate the pilot may have to listen first to the conditions for some twenty other airfields. If he is distracted at the crucial time he must wait until the information about all the others is transmitted before he again hears the values for the one he wants. In this way the cockpit workload is increased and drills may be interrupted. Although the broadcasts are in English, their quality is variable and in some cases may become almost incomprehensible. (*Editor's note*: Dr. Ruffell Smith's paper was illustrated with recordings of voice messages from pilots of different nationalities calling into the control tower at Heathrow Airport, London. It is unfortunate that the recordings cannot be reproduced in this book, because they illustrate dramatically the nature of the communication problem he discusses here.)

SOLUTION

Most of the communication needs and problems so far discussed can be met by the use of data links. These have many advantages. The messages are displayed visually, avoiding the errors due to short-term memory problems of nonpersistent audio displays. Since the visually presented messages are persistent and not time-constrained, they can be referred to at a time chosen by the pilot. In this way they do not add appreciably to peak workloads. Discrete addressing ensures that only those messages intended for a particular aircraft are displayed. In this way the pilot does not have to use some of his attention to listen for his own call-sign, or ensure that the channel is not in use before transmitting. With data links the actual transmission time is so compressed that only one-tenth the number of channels may be needed for the same number of aircraft. Finally, only a few relatively simple symbols are required, and these can be annotated in the language of the user.

Data links can be "general" or "special purpose." If all messages must be sent through one system only, this presupposes a full alpha numeric keyboard input and sophisticated readout, probably employing cathode ray tube techniques. However, if the majority of the traffic concerns a few special sets of information, single-purpose data links can be comparatively simple and cheap. Examples are links for air traffic control that may be used to direct aircraft by receiving aircraft position reports in machine language and by the symbolic display of instructions for turning and changing height with numerical displays for the desired heading, flight level, and airspeed (Ruffell Smith & Parker 1969).

Special systems for one-way transmission of meteorological information have also been devised. In these, the information from each airfield is encoded digitally and its transmissions preceded by a discrete label. The receiver is made selective to the label so that only the information about a particular airfield is displayed to the pilot, who is able to make the selection by setting up two switches, each having ten numbered positions. In this way information about the weather for any one of ninety-nine stations can be made available continuously in a simple format on a visual display.

To sum up, data links can eliminate most of the language problems associated with the use of voice communication for the operation of civil aircraft. They can recover the common language advantage inherent in the original digital system of W/T, lighten the load on short-term memory, eliminate most call-sign errors, and increase tenfold the capacity of radio channels.

REFERENCES

Ruffell Smith, H. P. Some human factors of aircraft accidents involving collision with high ground. *Journal of the Institute of Navigation*, 1968, **21**, 354–63.

Ruffell Smith, H. P., and Parker, B. D. The interaction of communication with cockpit workload and safety. Paper 2 in AGARD Advisory Report 19: *Aeromedical aspects of radio communication and flight safety*. Neuilly-sur-Seine, France: Advisory Group for Aeronautical Research and Development, North Atlantic Treaty Organization: December 1969.

FIFTEEN

A Survey of Driver Opinions in Three Countries

CARL GRAF HOYOS AND HEINZ-JÜRGEN LUTZE

The rapid growth of highway traffic in most Western countries has been accompanied by a steady increase in traffic accidents. Many steps have been, and are being, taken in a number of countries to reduce accident rates and to improve traffic safety. Improvements in highway systems and the construction of safer automobiles represent two areas in which efforts are being made to find effective countermeasures. Another approach is through the enactment of traffic legislation. In Germany, for example, a new traffic law was passed in 1971 that reformulated existing traffic regulations to conform more closely to current conditions. The passage of this law was accompanied by an extensive advertising campaign by the government to make drivers aware of the law, to increase its acceptability, and to improve drivers' attitudes.

In the area of driver education, experts are currently discussing the validity of the driver's license. In general, discussions have centered around such questions as the length of time that a driver's license, once issued, should remain valid; examination procedures; and the introduction of driver education into the school system. Until recently, however, the question of content in the curriculum of driving schools has very seldom been examined.

An analysis of the practice of driver training shows that most emphasis is currently placed on the proper handling of motor vehicles in routine

Carl Graf Hoyos, Technische Universität, München, Federal Republic of Germany; Heinz-Jürgen Lutze, Universität Regensburg, Federal Republic of Germany.

traffic situations. On the other hand, license examinations require knowledge about traffic signs, laws governing rights-of-way, and other legal matters. It is questionable, however, whether such knowledge is sufficient for safe driving. Presumably, there are many other areas that could be covered and that the driver would find much more beneficial.

Some of our current thoughts have come from interviews with drivers who have been driving for many years. These people were licensed ten or twenty years ago under completely different traffic conditions. They often have considerable driving experience without ever having become involved in a traffic accident. Yet many of these competent drivers would not be able to pass the theoretical part of a driver's examination today due to the fact that many traffic regulations have changed over the years, and the drivers have not received any formal instruction about those changes. This is true at least for the Federal Republic of Germany. The large amount of practical driving experience that many drivers have and the risks to which they expose themselves do not lend support to the idea that they have remained accident-free merely by chance. On the contrary, it seems plausible that safe driving involves more than information about road signs and traffic regulations, that is, the kind of knowledge that is required for licensing.

If our conjectures are correct, the question then arises about what sorts of knowledge and skill drivers should acquire in order to be successful and safe drivers. A good many teachers in driving schools are already aware of the inadequacies of present curricula. They try not only to prepare their students to pass the licensing examination but also to discuss problems which they, as experienced drivers, consider important for safe driving. There is considerable disagreement, however, about the relative importance of certain topics due to the personal interests and experiences of individual instructors. In our view, the curricula of driving schools and the examinations given by the license boards should be based on the results of investigations dealing with the knowledge and skills of people who have had considerable driving experience. We feel that this information should also be made available to younger drivers.

The goal of the investigation we report here was to compile a list of topics that are generally accepted as important and necessary for safe driving and that should be included in driver education courses. We were also interested in discovering whether the suggestions for improving driver education that were obtained from our study would be valid only in a particular country or whether similar results could be expected in other countries. This question is of some practical value because the members of the Common Market want to coordinate their traffic laws. To obtain some answers to this question, we conducted parallel investigations in England, France, and the Federal Republic of Germany.

METHOD

DEVELOPMENT OF THE INVENTORY

To prepare a list of topics dealing with driver knowledge, at least two approaches are available. First, one may ask experienced drivers, driving school instructors, and other experts about topics they consider to be relevant. This procedure would be very time-consuming and would result in a vast amount of redundant information. We decided, therefore, on another possibility and conducted a thorough study of the available literature as a basis for our investigation. Our sources were, for the most part, American textbooks on driver education. We also found raw material for the inventory in handbooks of driver training and in articles in automobile magazines and psychological journals. We took care not to exclude any topic that an author considered to be an integral part of safe driving. This resulted in an exceptionally long list of topics and statements.

We then reduced the number of entries by eliminating topics mentioned more than once and by consolidating problems that differed from others only in their wording. Further reduction came through the formation of categories that covered entire subject areas. The following requirements had to be met before we included a statement in the final list of relevant topics:

- Each topic had to include a new problem that had not been previously considered.
- The new topic could not be subordinate to any other problem unless it greatly increased the scope of the latter.
- If possible, the new topic had to be logically independent of the other topics, but at the same time its content had to be clearly definable.
- The topic had to deal with an extensive and important problem.

Care was also taken to formulate the topics as neutrally and as uniformly as possible in order to avoid the power of suggestion that could be exercised through the choice of words or through the scope and nature of the problem statement. Several wordings for each topic were worked out. These were then submitted to a panel of experts for discussion. From that we then arrived at a precise wording in its final form.

Following these procedures we ended up with 54 items in the final list of relevant topics, each representing an area of knowledge about traffic. The number 54 is somewhat arbitrary. On the one hand, we tried to establish as extensive a list as possible. On the other hand, we did not want the list to be so large as to be cumbersome. Under the circumstances, we cannot claim to have a complete list. Even so, it seems to us that we have considered the most important topics mentioned in the literature.

Problems in translation. In this paper we shall not discuss in detail problems of translating an inventory such as the one we used. However, we ought to mention a few topics because of particular translation difficulties that arose or because of differences in laws and regulations.

Item 13 (see Table 1) deals with intersections where traffic signs are not present to indicate the right-of-way. The rule is that a car approaching an unmarked intersection has to yield to traffic approaching from the right. There is no such regulation in Great Britain. As a result, on the British version of the questionnaire we changed the statement to the following: "Knowledge of the rules regarding right-of-way at junctions." We also had difficulty with the precise wording of this item in French.

Item 16 is concerned with defensive driving. From the number of question marks appearing on the returns it is clear that the concept of defensive driving is not a familiar one to many French drivers.

Item 51 is concerned with the identification of signposts by their different colors (in the German version). In Germany, road signs pointing to express-ways (*Autobahnen*) are blue; signs pointing to highways are yellow; and road signs for orientation within a city have black letters on a white background. Neither Great Britain nor France uses this method of differentiating roadsigns. The British and French translations refer, therefore, to the differentiation of roadsigns according to their shape.

Item 54 deals with registers for traffic offenders and level of fines for driving offences. On the advice of our French colleagues we eliminated this topic, because it is considered to be a contentious question for French drivers. Instead, we included the problem "Car radios and traffic safety" in the French inventory.

Table 1 gives the inventory in its entirety. Items are arranged in order of the average importance ratings assigned to each item by our German sample.

ADMINISTRATION OF THE INVENTORY

The individual items in the list were rated in three different ways. First, the evaluator had to rate his own knowledge about the subject; second, he was required to evaluate the importance of the subject for traffic safety; third, he was asked to judge how much knowledge about the subject he would expect to find among people with only little practical driving experience, that is, beginners.

The topics were compiled in a random order and each item was rated in one of the six categories shown in Table 2. As you can see, the categories are identical for the first and third questions, because in both instances knowledge is to be rated.

In addition, personal data, such as age, profession, and driving experience, were obtained from each of the informants.

Table 1. The 54 inventory items in all three translations. Ratings are mean importance ratings assigned to each item by the subjects in each sample

Item number	Item	Mean rating	Rank
1.	Richtiges Fahrverhalten beim Überholen	5.84	1
	Knowledge about overtaking	5.33	5.5
	Connaissances concernant le dépassement	5.50	7
2.	Kenntnis der Vorfahrtregelung	5.54	2
	Knowledge of the rules regarding right of way	4.84	17.5
	Réglementation concernant la priorité	5.58	3
3.	Wirkung von Alkohol auf das Fahrverhalten	5.51	3
	Effects of alcohol on driving	5.48	2
	Alcool et conduite	5.79	2
4.	Verhalten des Fahrzeugs beim Bremsen	5.49	4
	Reaction of the vehicle while braking	5.38	3
	Connaissances relatives à la tenue de route pendant le freinage	5.35	12
5.	Richtiges Verhalten bei Glatteis	5.44	5
	How to drive on icy roads	5.26	8.5
	Conduite par temps de verglas	5.38	10
6.	Sicherheitsabstand bei verschiedenen Geschwindigkeiten	5.42	6
	Safe distance between one's own car and the car in front at different speeds	5.51	1
	Variation des distances d'arrêt en fonction de la vitesse	5.52	5.5
7.	Notwendigkeit der Rücksichtnahme bei spielenden Kindern	5.40	7.5
	Need of caution in areas where children are playing	5.35	4
	Nécessité d'avoir regard aux enfants qui jouent	5.13	17
8.	Richtiges Fahrverhalten und angemessene Geschwindigkeit bei verringerter Bodenhaftung	5.40	7.5
	Driving and speeding in case of reduced grip on the road surface	5.33	5.5
	Conduite et adjustement de la vitesse sur chaussée glissante	5.42	8
9.	Kenntnis der Verkehrszeichen	5.30	9
	Knowledge of traffic signs	4.87	16
	Connaissance de la signalisation routière	5.52	5.5
10.	Richtiges Fahrverhalten bei ungünstigen Sichtverhältnissen	5.26	10
	Driving under bad conditions of visibility	5.26	8.5
	Conduite lorsque la visibilité est réduite	5.38	11
11.	Richtiges Durchfahren von Kurven	5.14	11.5
	How to drive round bends	5.05	12
	Conduite en virage	5.00	21

Table 1 (continued)

Item number	Item	Mean rating	Rank
12.	Richtiges Fahrverhalten bei Sturm und bei Windböen	5.14	11.5
	Driving in a gale or gusts	4.22	34
	Conduite par mauvais temps (pluie, vent)	5.15	16
13.	Kenntnis der besonderen Bedingungen, die an bestimmten Kreuzungen und Einmündungen häufig zu Vorfahrtfehlern führen	5.12	13
	Knowledge of the rules regarding right of way at junctions	5.00	13
	Franchissement des intersections (signalisation, priorités, fautes de priorité)	5.81	1
14.	Richtiges Verhalten nach einem Unfall	5.09	14
	What to do after an accident	4.37	26
	Connaissance de ce qu'il faut faire en cas d'accident	5.10	18
15.	Kenntnisse in Erster Hilfe	5.05	15
	Knowledge of first aid	3.66	41
	Secourisme routier	4.71	31
16.	Defensives Fahren und Verkehrssicherheit	4.93	16
	Defensive driving and road safety	4.78	20
	Conduite défensive et sécurité	4.35	37
17.	Bedeutung der Reaktionszeit für das Fahrverhalten	4.91	17
	Significance of reaction-time in driving	4.90	14
	Rôle du temps de réaction du conducteur	5.25	13
18.	Strassenbelag und Länge des Bremsweges	4.88	18
	Dependence of braking distance on road surface	5.20	10.5
	Nature du revêtement routier et distances de freinage	4.77	30
19.	Besonderheiten beim Fahren im Winter	4.86	19
	Particular points to notice when driving in winter	4.29	31.5
	Conduite en hiver	4.40	36
20.	Richtiges Ein- und Ausfädeln bei Bundesstrassen und Autobahnen	4.81	21
	Entering and exiting from main roads and motorways	5.20	10.5
	Comment entrer sur les autoroutes et en sortir	4.83	27
21.	Bedeutung von Sehschärfe, Farbensehen, Nachtsehen und Blendung für das Fahrverhalten	4.81	21
	Importance of visual acuity, colour discrimination, dark adaptation, and headlight-dazzle for driving	4.23	33
	Rôle de l'acuité visuelle, de la vision des couleurs et de la vision nocturne	5.08	19
22.	Beachtung des sogenannten toten Winkels	4.81	21
	Need to be aware of the so-called blind spot of restricted vision	4.84	17.5
	Problèmes de visibilité posés par des "angles morts"	4.79	28.5

Table 1 (continued)

Item number	Item	Mean rating	Rank
23.	Verkehrsverhalten älterer Fussgänger	4.74	23.5
	Behaviour of elderly pedestrians	4.61	24
	Comportement des piétons âgés	4.69	32
24.	Probleme, die mit dem Kolonnenspringen zusammenhängen	4.74	23.5
	Problems concerned with overtaking in heavy traffic	5.27	7
	Dépassement dans la circulation dense	4.88	25
25.	Richtiges Verhalten beim Passieren von Haltestellen	4.72	25
	How to drive near bus-stops	3.99	36
	Passage à proximité des arrêts d'autobus	4.25	39.5
26.	Vorausschätzung des Verhaltens ungeduldiger oder eiliger Fussgänger	4.67	26
	Ability to predict the behaviour of pedestrians who are impatient or in a hurry	4.66	22.5
	Capacité de prévoir ce que vont faire les piétons pressés	4.79	28.5
27.	Bedingungen, die die Einschätzung von Geschwindigkeit beeinflussen	4.58	27.5
	Conditions which influence the estimation of speed	4.59	25
	Conditions qui influencent l'estimation des vitesses	4.56	35
28.	Ermüdung und Monotonie beim Autofahren, erste Anzeichen und Möglichkeiten der Abhilfe	4.58	27.5
	Fatigue and monotony while driving, their first signs and how to overcome them	4.81	19
	Signes de fatigue et d'endormissement pouvant avertir le conducteur de son état et moyens d'y remédier	5.17	15
29.	Beachtung und Vorausschätzung des Verhaltens von Rad-und Mopedfahrern	4.11	29
	Ability to estimate and predict behaviour of cyclists and motorcyclists	4.66	22.5
	Capacité a prévoir ce que vont faire les cyclistes et les cyclomotoristes	4.92	22.5
30.	Fragen der Beleuchtung beim Kraftfahrzeug	4.47	30
	Questions concerned with the lights on a motor vehicle (front lights, rear lights, direction indicators etc.)	3.90	37
	Eclairage du véhicule	5.38	9
31.	Probleme beim Fahren unter Zeitdruck	4.44	31
	Problems involved when driving in a hurry	4.73	21
	Influence de la hâte (conduite quand on est pressé)	4.85	26
32.	Richtiges Verhalten beim Bemerken von Fahrzeugen mit Blaulicht	4.42	32
	Correct reaction when one sees an emergency vehicle (for example, an ambulance)	4.29	31.5
	Comportement en présence d'un véhicule prioritaire (ambulance, police, pompiers)	5.19	14

Table 1 (continued)

Item number	Item	Mean rating	Rank
33.	Richtiges Verhalten bei Autopannen	4.26	33
	What to do when one's car breaks down	3.15	49
	Connaissances utiles en cas de panne	3.31	53.5
34.	Besonderheiten beim Fahren im Stadtverkehr, auf Landstrassen und Autobahnen	4.21	34
	Special problems to notice when driving in urban areas, in rural areas, and on motorways	4.36	27.5
	Différences entre la conduite en ville, sur la route et sur autoroute	4.67	33
35.	Notwendigkeit der Einhaltung vorgeschriebener Geschwindigkeiten	4.19	35
	Need to observe the speed limits	4.33	29.5
	Nécessité de respecter les limitations de vitesse	4.92	22.5
36.	Beobachtung von Besonderheiten beim voraus-fahrenden Fahrzeug	4.16	36
	Observing everything unusual about the vehicle in front	3.86	39
	Prévision des manoeuvres de la voiture qu'on suit	5.02	20
37.	Notwendigkeit und Problematik der Geschwindig-keitsbegrenzung	4.00	37
	Need for speed limits and the problems involved	4.13	35
	Problèmes posés par la limitation de la vitesse et la nécessité de celle-ci	4.23	41
38.	Kenntnis der wichtigsten Paragraphen der Strassenverkehrsordnung und des Strassen-verkehrsgesetzes	3.84	38
	Knowledge of the most important rules of the road and the traffic regulations	4.88	15
	Connaissances des points importants du Code de la Route	5.56	4
39.	Einfluss von Stimmungen und Launen auf das Fahrverhalten	3.81	39
	Influence of moods on one's driving	4.33	29.5
	Influence de la joie, de la colère, de la tristesse (ennuis, soucis) sur le comportement des conducteurs	4.25	39.5
40.	Besonderheiten beim Fahren im Gebirge	3.80	40
	Particular problems when driving in the mountains	3.22	47.5
	Conduite en montagne	4.00	44
41.	Einfluss des Lebensalters auf das Fahrverhalten	3.63	41
	Influence of the driver's age on driving ability and behaviour	3.41	45
	Influence de l'âge sur le comportement des usagers	3.73	47
42.	Vor- und Nachteile verschiedener Reifenarten	3.58	42
	Advantages and disadvantages of different types of tires	3.84	40

Table 1 (continued)

Item number	Item	Mean rating	Rank
	Qualités et défauts des différents types de pneumatiques	3.85	45
43.	Gesichtspunkte, die mit der Beladung des Kraftfahrzeugs in Zusammenhang stehen	3.54	43.5
	Problems concerning the loading of a vehicle with luggage, etc.	3.40	46
	Chargement du véhicule (maximum, répartition, arrimage, visibilité, longueur, etc.)	4.58	34
44.	Vor- und Nachteile der Sicherheitsgurte	3.54	43.5
	Advantages and disadvantages of safety-belts	4.36	27.5
	Avantages et désavantages des ceintures de sécurité	4.29	38
45.	Besondere Fahreigenschaften von Lastwagen und Spezialfahrzeugen	3.51	45
	Particular characteristics of lorries and special vehicles on the road (tanks, car transporters etc.)	3.52	42
	Comportements dans la circulation particuliers aux véhicules lourds	4.08	34
46.	Vorbereitung längerer Fahrten	3.40	46
	Planning of longer trips	3.07	50
	Préparation d'un voyage	3.54	48
47.	Auswirkung der Verkehrsregelung durch Ampeln und Vorfahrtszeichen auf den Verkehrsfluss	3.35	47
	Effect of traffic lights and right-of-way signs on the traffic flow	3.47	44
	Influence sur l'écoulement du trafic des feux et des signaux	3.81	46
48.	Besonderheiten bei Fahrten ins Ausland	3.26	48
	Special points to consider when driving in foreign countries	3.50	43
	Conduits dans les pays étrangers	4.15	42
49.	Rauchen während der Fahrt	3.12	49
	Smoking behind the wheel	2.95	52
	Influence du fait de fumer sur la conduite	3.50	49
50.	Technische Kenntnisse über Motor, Lenkung und Bremsen	3.02	50.5
	Technical knowledge about engine, steering, and brakes	3.02	51
	Connaissances techniques (moteur, transmission, frein)	3.33	51.5
51.	Unterscheidbarkeit von Wegeweisern an Hand ihrer Farbe	3.02	50.5
	Identification of signposts by their different shapes	3.87	38
	Différentiation des panneaux de signalisation selon la forme	4.90	24

Table 1 (continued)

Item number	Item	Mean rating	Rank
52.	Einteilung und genaue Abgrenzung der Führerscheinklassen	2.88	52
	Exact classification of the different types of driving licences	1.81	54
	Véhicules pouvant être conduits avec les diverses catégories de permis	3.48	50
53.	Versicherungsrechtliche Bestimmungen im Zusammenhang mit dem Kraftfahrzeug	2.79	53
	Insurance regulations concerning motor vehicles	3.22	47.5
	Réglementation concernant les assurances	3.31	53.5
54.	Verkehrssünderkartei und Bussgeldkatalog	2.07	54
	Registers for traffic offenders and level of fines for driving offences	1.87	53
	Autoradio et sécurité	3.33	51.5

SAMPLES

Federal Republic of Germany. The inventory was given to different groups of drivers, including truck drivers, bus drivers, private drivers, and beginners. To facilitate comparisons among the three countries, we tried to get only groups of drivers that could be designated "average" with respect to experience and other characteristics. For this reason, we selected in Germany a sample from a group of individuals taking courses in administration and economics that are given to clerks and Public Service employees who are trying to qualify for higher positions. These people were all willing to cooperate, were well motivated, and had the necessary education to understand the inventory. On the other hand, none of them had a university education.

Forty-three out of the original group of seventy male drivers to whom the inventory was sent returned their inventories completely filled out (return rate = 61 percent). The participants averaged thirty years in age and had had their driver's licenses for approximately nine years. They obtained their driving experience mainly in rural areas or in small communities.

France and Great Britain. The results of the survey in Germany were to be compared with corresponding surveys in France and Great Britain. We, therefore, contacted colleagues[1] in these countries and asked them to correct the translations of the inventory and then to help us by conducting a

[1]We thank Director M. Roche from the Centre de Recherches et Applications de la Prévention Routière in Linas-Montlhéry and Dr. A. B. Clayton of the Department of Transportation and Environmental Planning, Road Accident Research Unit of the University of Birmingham, for their assistance.

Table 2. Frames of reference and response categories used by our subjects in rating each item in the questionnaire

Frame of reference	Response categories					
	1	2	3	4	5	6
Subject's assessment of his own knowledge	ungenügend	kaum ausreichend	ausreichend	zufriedenstellend	gut	ausgezeichnet
	inadequate	hardly adequate	adequate	satisfactory	good	excellent
	insuffisant	à peine suffisant	suffisant	satisfaisant	bon	excellent
Importance for road safety	nicht wichtig	weniger wichtig	durchaus noch wichtig	wichtig	sehr wichtig	äusserst wichtig
	unimportant	not very important	fairly important	important	very important	extremely important
	pas important	moins important	d'une certaine importance	important	très important	extremement important
Assessment of beginner's knowledge	ungenügend	kaum ausreichend	ausreichend	zufriedenstellend	gut	ausgezeichnet
	inadequate	hardly adequate	adequate	satisfactory	good	excellent
	insuffisant	à peine suffisant	suffisant	satisfaisant	bon	excellent

similar study in their respective countries. They distributed copies of the inventory to samples of drivers comparable to the group of Germans and then returned the inventories to us.

Inventories were distributed to 120 individuals in different parts of France. Most of them were industrial employees and were taking courses in economics. Their average age was thirty-four. With respect to education and socioeconomic status, they correspond quite well to our sample in Germany. Fifty individuals returned the inventory (return rate = 42 percent). Two of those returned were incomplete. In all, forty-eight could be evaluated.

In Great Britain the procedure for distributing the inventory was somewhat different from that in Germany and France. Contacts in the West Midlands were asked to supply addresses of drivers who were presumably able to participate in the study. Restrictions were made only with regard to age and sex. At present 89 out of a sample of 120 who were contacted have returned the inventories (return rate = 74 percent). Since 3 inventories were not completely filled out, 86 were evaluated.

In the British sample, which included employees, laborers, teachers, and students, there was greater variability in education and vocation. This sample is therefore more representative of the population as a whole, but at the same time not as comparable to the other samples because of its diversity. Some characteristics of the sample groups are summarized in Table 3.

Differences in average age between the French sample on the one hand and the English and German samples on the other are significant at the 1 percent level. On the average, the French sample has more driving experience than the German sample, and this difference is significant at the 5 percent level. However, we do not believe that these differences are serious enough to invalidate other comparisons among our samples. Our primary aim was to get middle-aged individuals not older than fifty who had driven regularly for at least three years and had driven at least 64,400 kilometers (40,000 miles). These conditions have obviously been met in our samples. In short our subjects are driving "experts." They are not representative of drivers in general.

Table 3. Age and driving experience (in years) of the three samples of drivers

		Age			Driving experience		
Country	N	Mean	Range	S.D.	Mean	Range	S.D.
Germany	43	30.0	23–42	2.1	9.1	3–15	3.1
Great Britain	86	29.8	23–46	4.9	10.4	4–28	4.5
France	48	34.7	24–49	7.0	11.4	5–22	4.8

RESULTS

Table 1 gives for each item in the inventory its mean importance rating and the rank order of the mean rating computed separately for the three samples of drivers. Although the table shows some striking agreements among the ratings and ranks for the three samples (see, for example, items 3, 8, 10, 17, 42, 47, 50, 52, and 54), there are also some large differences. The most impressive disagreement is that for item 38 which is ranked fourth (and so of major importance) by French drivers, but thirty-eighth (and so of much less than average importance) by German drivers. French drivers evidently regard it as important to know the rules of the road; German drivers regard it as much less important, while British drivers fall between the two.

The second largest difference is for item 51 which, once again, is regarded as of more than average importance (rank of 24) by French drivers and of almost trivial importance (rank of 50.5) by the Germans. Recall, however, our earlier comment that the wording of this item is not the same in all three languages.

Other large differences, in decreasing order of size, are for items 15, 12, 16, 30, 33, 24, 32, and 19. Some of the differences—for example, the great importance attached to knowledge of first aid, winter driving, and driving under windy or gusty conditions by the Germans—are relatively easy to account for. A knowledge of first aid is now required by German licensing examinations, and Germans, much more frequently than the British, have to drive under dangerous winter and windy conditions. On the whole, however, the discrepancies among the ranks for individual items do not form consistent patterns that reflect major differences in driver opinions among the three national groups.

In Table 4 means and standards deviations have been computed for all the items in the inventory, but have been computed separately for each of the three national samples and for each of the three frames of reference used in rating the items.

The small differences among countries in Table 4 could be due to differences in the interpretation of individual items (Table 1) or of response categories (Table 2). That is, the several items and scale values were probably not identical in the English, French, and German versions of the questionnaire. On the other hand, it is also possible that some or all of the differences observed among the means were determined by differences among the samples themselves.

A better indication of the general agreement among the three samples is provided by coefficients of concordance (W) for the rankings of the mean ratings made for each item in each frame of reference (as, for example, among the ranks for importance in Table 1). The results are in Table 5. If we take into account the numerous differences in traffic laws, driving habits, and road conditions among the several countries, as well as unavoidable inaccuracies in sampling techniques, the agreement among the three groups is amazing. Despite this general conformity, there were, as we have already pointed out, some differences from country to country. Still other differences will be indicated in the discussion that follows.

DIFFERENCES BETWEEN OPINIONS ABOUT EXPERIENCED AND INEXPERIENCED DRIVERS

The differences among the columns in Table 4 are far greater than the mean differences among the German, English, and French samples. In all three samples, when asked to judge their own knowledge of a subject, raters tended to use categories at the upper end of the scale (excellent, good, satisfactory), whereas for the beginner they used categories toward the bottom (sufficient, hardly sufficient, insufficient). This corresponds quite well with conventional stereotypes about beginning drivers. The most important exceptions are shown in Table 6. These exceptions are striking in view of the much lower average ratings given to estimates of beginners' knowledge as compared with the mean ratings experienced drivers give themselves (Table 4).

Table 6 shows that the three samples are in almost complete agreement. Beginners are judged to know more about those topics, namely road signs and traffic regulations, that are dealt with directly in driving schools and that are tested in the licensing examination. In Germany, this was also true of first aid, because people applying for a driver's license in that country must now have a knowledge of this subject. Since this is a recent development, beginning drivers are judged to be better informed about this subject than are experienced drivers.

IMPORTANT TOPICS ABOUT WHICH DRIVERS NEED MORE INFORMATION

Other differences of some interest concern those subject areas to which raters attach great importance, but about which they claim little knowledge

Table 4. Arithmetic means and standard deviations of the ratings

Country	Frame of reference: Own knowledge	Importance	Knowledge of beginners
Germany	M = 4.5 S.D. = 1.2	M = 4.4 S.D. = 1.3	M = 3.6 S.D. = 1.4
Great Britain	M = 4.4 S.D. = 1.3	M = 4.3 S.D. = 1.4	M = 2.9 S.D. = 1.4
France	M = 4.5 S.D. = 1.3	M = 4.7 S.D. = 1.2	M = 3.0 S.D. = 1.5

Table 5. Coefficients of concordance among the samples

Frame of reference	W
Assessment of subjective knowledge	0.82
Importance for road safety	0.88
Assessment of beginner's knowledge	0.87

(Table 7). Only two such topics are common to all three countries: (1) knowledge about first aid and (2) safe following distances at different speeds. The German and French drivers apparently need to know more about (1) what to do after an accident and (2) how vehicles behave during braking operations. French and British drivers, on the other hand, feel the need for more information about driving under icy conditions. German drivers are unique in judging that they lack sufficient information about the relationship between braking distances and road surfaces. There are two such unique items for British drivers: (1) information about driving in foreign countries and (2) driving under conditions of reduced traction.

Table 6. Areas with which beginners are judged to be more knowledgeable than experienced drivers.[a]

Country	Item	Knowledge of beginners	Own knowledge	Difference
Great Britain	Identification of signposts (51)	5.0	4.3	0.7
	Knowledge of the rules (13)	5.0	4.6	0.4
	Knowledge of traffic signs (9)	5.2	5.0	0.2
	Need to observe the speed limits (35)	4.6	4.5	0.1
France	Knowledge of the rules (13)	5.4	5.0	0.4
	Identification of signposts (5)	5.0	4.6	0.4
	Knowledge of traffic signs (9)	5.3	5.0	0.3
Germany	Knowledge of first aid (15)	4.3	2.7	1.6
	Knowledge of the rules (13)	4.9	3.9	1.0
	Registers for traffic offenders (54)	4.0	3.4	0.6
	Knowledge of traffic signs (9)	5.4	5.1	0.3
	Identification of signposts (51)	4.6	4.5	0.1

[a]Numbers in parentheses are the item numbers given in Table 1.

Table 7. Topics showing large differences between mean ratings of importance and mean ratings of self-knowledge[a]

Country	Topic	Importance		Knowledge		Difference between mean ratings
		Mean rating	Rank	Mean rating	Rank	
Germany	Knowledge of first aid (15)	5.0	15	2.7	54	2.3
	What to do after an accident (14)	5.1	14	4.1	44	1.0
	Safe distance at different speeds (6)	5.4	6	4.5	31	0.9
	Braking distance and road surface (18)	4.9	18	4.2	42	0.7
	Reaction of the vehicle while braking (4)	5.5	4	4.9	17	0 6
Great Britain	Driving in foreign countries (48)	3.5	43	2.4	53	1.1
	Knowledge of first aid (15)	3.7	41	2.6	52	1.1
	How to drive on icy roads (5)	5.3	8.5	4.4	36	0.9
	Safe distance at different speeds (6)	5.5	1	4.7	19	0.8
	Driving in case of reduced traction (8)	5.3	5	4.5	27	0.8
France	Knowledge of first aid (15)	4.7	31	2.9	54	1.8
	How to drive on icy roads (5)	5.4	10	4.0	44	1.4
	What to do after an accident (14)	5.1	18	3.8	50	1.3
	Safe distance at different speeds (6)	5.5	5	4.4	31	1.1
	Reaction of the vehicle while braking (4)	5.4	12	4.5	28	0.9

[a]Numbers in parentheses are the item numbers given in Table 1.

Once again, several of these differences are understandable in view of national differences in licensing requirements and driving conditions. We have already commented on the knowledge of first aid that is now required for licensing in Germany. The British interest in knowing more about how to drive abroad is undoubtedly a reflection of the fact that the British drive on the left side of the road, in contrast to what applies throughout the European continent. Similarly, the British and French probably feel the need to know more about driving under icy conditions because they encounter such conditions less frequently than do the Germans.

CONCLUSION

This study has sampled the opinions of experienced drivers in three countries: Germany, Great Britain, and France. Although there are some differences among the three groups of drivers, the general picture is more one of agreement than of disagreement. It is important to add, however, that despite differences in language, customs, vehicles, and road conditions, the people in these three countries are probably much more alike in their driving habits than one might expect to find in other parts of the world. Nonetheless, it seems clear that driving instruction in the three countries should differ in some details to satisfy the special requirements of drivers in those countries. Our findings also contain some implications for the design of vehicles, signs, and highways in the several countries sampled.

SIXTEEN

Human Factors in Town Planning and Housing Design

BARUCH GIVONI

In recent years human aspects of urbanization in general, and of housing as a particular facet of it, have received very wide coverage in the professional literature dealing with urban problems. However, a relatively small part of that literature has been concerned with the relationship between human factors and the detailed *physical planning* of urban environments. This paper discusses some of the sociocultural factors encountered in town planning and housing design.

Over centuries past, climatic and sociocultural factors were among the main forces that shaped the indigenous forms of human settlements and housing in different regions of the world. By adapting to climatic conditions while adjusting to sociocultural specific needs, housing types were evolved suitable to the way of life of the inhabitants. In general, such housing provided the best indoor climate attainable given the local climate and the technology available.

Technological and social conditions do not, however, remain constant, and in recent decades they have been changing at an accelerated rate. New building and urban utilities technologies now make possible types of construction and urban densities unattainable heretofore. At the same time, the same socioeconomic forces that have promoted the mass urbanization process have also caused a sharp increase in levels of expectation. Housing conditions that were accepted as natural in the past are no longer tolerated. These rapid changes have resulted in new forms of human settlements and housing types being developed over a short period, without the necessary time for "natural" adaptation to (a) the climatic conditions prevailing in different regions and (b) changing sociocultural needs.

Baruch Givoni, Building Research Station, Technion, Haifa, Israel.

Meanwhile, the professions of architecture and town planning have become "internationalized" as a result of the very fast communication now possible between professionals all over the world. This, in turn, has been responsible for the tendency to imitate the latest fashions, disregarding local climatic and sociocultural conditions. The situation is further complicated in developing countries, because in most cases the architects and planners come from, or were trained in, developed countries with completely different climatic and sociocultural backgrounds.

As a result, one can hear frequent complaints about indoor climatic conditions, which in many types of new housing are worse than those found in traditional houses. Furthermore, various features of the physical planning of housing and its environment are found frequently unsuitable to the living patterns and special sociocultural needs of inhabitants with different national or sociological backgrounds. When urbanization is accompanied by mass public housing programs, where hundreds or thousands of similar dwelling units are constructed, such mistakes can have severe consequences.

Clearly, there is a need to develop methods of preevaluating the suitability of given planning schemes to meet human requirements. Such methods should be specific for the different types of human requirements. As an example, a procedure has been developed for evaluating housing designs from the viewpoint of their thermal comfort (Givoni 1969). This paper will mention some implications of sociocultural factors for thermal conditions.

The general procedure for insuring compatibility between sociocultural needs and planning features should involve the following stages:

- Identification of the sociocultural variables relevant to planning and explicit description of the corresponding human needs.
- Identification of the planning elements that may affect a given sociocultural variable and the nature of their effects.
- General analysis of the probable effects of different solutions of a given planning element on the corresponding sociocultural needs.
- Examination of the adequacy of a proposed planning solution, using methods and procedures appropriate for evaluating the impact of that solution on different sociocultural variables.

Some sociocultural variables that should be considered in planning a given urban or housing project for a particular population are:

1. Attitudes about privacy
2. Family structure
3. Role of women in the family and in society
4. Child-rearing patterns
5. Recreation patterns
6. Entertainment patterns
7. Shopping habits

8. Cooking and eating habits
9. Social interaction patterns
10. Socioeconomic and residential mobility

ATTITUDES ABOUT PRIVACY

Attitudes about privacy differ greatly among peoples of different cultures. Hall (1966) describes differences among Germans, Englishmen, and Americans in their attitudes and customs regarding privacy and their reliance on architectural features to achieve it. He observes that although Germans are much more appreciative of privacy than Americans, both use physical barriers, such as the closing of doors, whenever privacy is wanted. Englishmen, according to Hall, rely on subtle cues that signal one's desire for privacy. Other persons are supposed to recognize and to respond to these cues, even in the absence of physical barriers.

Much wider differences exist in attitudes about privacy among societies with different cultural backgrounds. African and Asian families attach high value to the provision of privacy, even within the house. Thus, Datta and Lal (1968) state that the tradition in India requires privacy for young ladies, even within a dwelling. Gupta (1968) has noted that in India the provision of privacy for women inside the home calls for the visual separation of the housewife's working areas from the living area.

Different attitudes toward privacy are reflected in different planning and design solutions. While visiting housing projects in Den Haag during the evening, I was surprised to observe buildings with very large windows facing the main street without any curtains. Anyone in the street could see everything going on in these illuminated rooms. On the other hand, in Ghana, I have seen many houses with windows facing the street permanently shut by solid wooden shutters, nailed to the walls, in spite of the intolerable heat inside these houses. It was explained to me that in recent years there has been an increase in the number of strangers from nearby cities coming to the villages. Consequently, the windows have been shut off to provide privacy and security from persons walking in the street. This practice demonstrates that when sociocultural needs are in conflict with comfort and convenience, the latter may be sacrificed to insure the former. A point that should be made in this connection is that, with our present state of knowledge on building climatology, it is possible to design openings that would satisfy the need for security and privacy and at the same time provide adequate ventilation.

Conflicts between planning details and needs for security and privacy have been reported in various studies. Datta and Lal (1968) have commented that in India, open planning and one-room low-cost apartments do not provide adequate privacy. Even visual contact through windows facing one another too closely leads to feelings of lack of privacy. Atkinson (1959) has observed that in Africa large rooms are frequently divided into smaller cubicles, either by curtains or by partitions, to provide a private personal space for the members of the family, even if it is very small.

PLANNING FEATURES AFFECTING PRIVACY

Urban density. Differences in attitudes about privacy and the readiness to sacrifice other amenities to achieve it are reflected by the different concepts regarding what constitutes a "tolerable" density. Thus in Hong Kong the net density of the housing built by the housing authority reaches 3,000 persons per acre (Parry & Stuber 1969), and the inhabitants manage to adjust to it. Such a density would be considered beyond the tolerable limit in most other societies, even for the poorest segment of the population.

It is commonly assumed that higher urban density, for example, more dwellings per unit area of the city, implies less privacy. In reality the actual privacy of inhabitants inside their dwelling depends greatly on the details of the urban planning and the house, and the way in which a given density is achieved. For example, visual and auditory privacy from strangers outside the dwelling depends on the distance between windows of different dwellings and on their relative heights and orientations. For a given density, more privacy will be provided by higher buildings far apart than by lower buildings closer together.

Distance of buildings from the street. This factor is of importance mainly for the inhabitants on ground floors. As the distance of a building from a street becomes smaller, less privacy may be provided from people walking in the street. This effect can be reduced, or even eliminated, by providing a separation between the street and the building, that is, a fence or wall.

Distances between buildings. In general, increasing the distance between buildings provides more visual and auditory privacy. When the distances are small, however, it is possible to increase visual, and to some degree also auditory, privacy by specially designed windows, and by changing the orientation of the windows relative to the façade.

Type of house. Single-family houses generally provide more privacy than do multifamily buildings. This is so because both the entrance to and the area around a single-family house are under the direct control of the occupant. These advantages of single-family houses diminish as urban densities increase and the distances between buildings decrease.

Row houses with separate, individual entrances to the dwelling units are considered next best to single-family houses from the privacy viewpoint. Feelings of privacy are enhanced if row houses contain front and back yards.

In the case of multifamily buildings, the feeling of privacy is diminished during entry and exit. However, the privacy inside a dwelling unit depends on the details of its design and construction and on the distances between buildings. With proper design, multifamily buildings may be even better than single houses built too closely together.

The number of rooms in the dwelling. Several details in the organization of the dwelling area affect the privacy of the individual members of the family. Among them is the number of separate rooms into which the dwelling is divided. When everyone in the family has a room of his own that can be closed off whenever he desires it, privacy can be secured by all the family. However, with a given total space, which in the case of low- and medium-cost housing is not large, subdivision into more rooms means a smaller average area for each room.

A room can be designed in such a way that it will satisfy basic needs, such as sleeping, reading and writing, and some storage, with an overall area of 4 to 5 sq m. In this case, privacy is achieved at the expense of spaciousness. In addition, the organization of the internal space becomes more difficult as the number of rooms increases. The internal circulation becomes more complex and requires a larger portion of the total floor area, and the allocation of external walls to the different rooms becomes more difficult.

To sum up, decisions about the number of rooms into which a total area is divided usually involve a conflict of considerations, for privacy on the one hand, and for spaciousness and more efficient organization of the dwelling area on the other.

"Open" or "closed" internal planning. The architectural character of the internal planning has great influence on the interference of the different activities taking place in the house, and on the privacy inside it. "Open" planning, in which the living zone is frequently connected visually with the kitchen, with study areas, and sometimes also with the bedrooms, reduces greatly the privacy of the family members. But such planning can provide a feeling of spaciousness, even in dwellings of relatively small area.

On the other hand, formal "closed" planning, in which there are separate rooms for the various functions of the dwelling, is more satisfactory from the viewpoint of privacy but of duller architectural expression.

Acoustical quality of the building. Noise creates irritation, particularly when its source is from neighbors. For this reason the acoustical quality of the neighborhood and buildings, that is, the attenuation of sounds through walls and floors and minimizing noise entry through windows, is of great importance not only in securing privacy but also in preventing unnecessary tension between neighbors.

From the privacy point of view, it is sometimes equally important to avoid the feeling that what is going on in one's home can be heard by his neighbors. For some people, the knowledge that they might be heard by their neighbors constitutes a big restriction on the freedom of their activities at home.

Different cultures differ greatly in the loudness of their living and their attitudes to noise. What is taken as a normal sound level in one culture might be considered as annoying and even intolerable noise by another. Social tension may develop when persons with different attitudes to noise live in close proximity.

In order to ensure an adequately low noise level and to avoid the transmission of conversations between housing units, the acoustical "connections" should be carefully designed. For example, the amount of separation and the angle between windows determines to what extent neighbors will or will not be heard. Dividing walls and floors between units should be adequately insulated. Internal walls have to allow for different noise-generating activities in any room, without interfering with the privacy of the other rooms. Another important source of sound propagation detrimental to privacy is the vertical ventilation ducts connecting water closets or toilets and bathrooms. Utility fixtures such as showers or water faucets should not be affixed to dividing walls between units. Doing so imposes serious limitations on privacy both for anyone who doesn't dare to take a bath at night and for his neighbor who has to hear it if he does.

Another problem may be posed in multifamily buildings by disturbing noise coming from common entrances, corridors, and internal courtyards. For example, a recent multifamily dwelling built in Eilat, Israel, which presents many new functional and architectural features aimed at obtaining improved indoor climate conditions, fails to ensure adequate "acoustical" privacy inside each unit. This happens because the covered courtyard which gives access to the units is at the same time a source of noise, magnified by the reflective surfaces which enclose the space. At the same time, noise created within the apartments may be heard in the courtyard through the openings of each unit that face the courtyard. This clearly shows that architectural innovations have to take into account all the parameters affecting comfort and satisfaction in the home and not solve some of them at the expense of others.

ROLE OF WOMEN

The role of women in the family and in society has important consequences for the design of homes. In the first place, monogamous or polygamous societies have different requirements. In the latter the house needs to be subdivided into semiautonomous subunits.

In societies in which women have a more restricted status than men, usually staying at home, shielded from the view of strangers, the design of the house needs to provide adequate privacy from the outside and a clear separation between the living quarters on one side and the bedrooms, kitchen, and utility rooms on the other. In those societies where the wife participates actively in receiving guests, the relation between kitchen and living room should be a close one, enabling the housewife to attend the guests and to participate in conversation at the same time.

Women who work outside their homes for a substantial part of the day are pressed harder for time, because they are still ordinarily responsible for the household work. In societies where married working women are prevalent, there is a need to provide facilities that will ease their load. On the neighborhood scale, kindergartens and day-care centers need to be provided for preschool children, and eating arrangements for school-age children. Public transportation connecting residential and the work areas, easily accessible at both ends, is also very helpful in many such communities. Even when a family has a private car, the car is usually used by the husband.

Planning must also have as a goal the saving of time by rationalizing shopping facilities (aggregating shops in shopping centers, for example) and by rationalizing the internal design of the home so as to facilitate home care. With respect to the latter, kitchens should be big enough to incorporate modern household equipment, and a storage place should be available for weekly purchases.

Women who do not work outside need more space at home for extra tasks (for example, sewing), and for the care of children.

FAMILY STRUCTURE

Great variations in family structure exist between different cultures and socioeconomic classes. The range is from the many-children extended family to the nucleated (basic) family with only 1 to 3 children. Family structure should have a prominent influence on planning features, both at the urban and residential levels. The main parameters of family structure which affect planning are the type of household, nucleated or extended family, and the "typical" number of children.

In societies with extended families, where several generations live together and married couples and their children stay with their parents, the house has to be an expandable compound so that new quarters can be added as the family grows. Land allocation and basic design features should permit such future expansion.

In several societies where extended families are prevalent, the indigenous housing type has as its dominant feature a central open courtyard around which the living units of the individual basic families are built. Besides offering many advantages from the climatic viewpoint in the hot season, such a design permits expansion of the compound as children grow older and get married and as more children are born.

The importance of accommodating more than a single basic family in the household residence, for the conditions prevailing in India, has been stressed by Gupta (1968). The needs of the extended family call for a large number of rooms within the housing compound. When the total building area is restricted for economic reasons, a house having more small rooms is more suitable than one having larger but fewer rooms with the same total area.

Similar housing compounds have also been developed in societies with polygamous families, where each of the women can have her own quarters to accommodate her and her children, within the same household.

The size of the family at a given socioeconomic level determines the amount of space that needs to be allocated to belongings and thus influences the total floor area needed. The relationship is not proportional, however, because children and parents differ in their space requirements. Hence the composition of the family affects the interrelationship between different functional areas and the details of house design.

In rural societies where land is abundant and houses are usually of one or two stories only, the house can be extended as the family grows and can be adjusted to its changing needs. Urbanization brings greater pressure on land and a shift to high density, multistory developments. The freedom of the family to adapt a house to its changing composition and needs is lost. As a result, the number of children and their ages influences residential preferences in urban areas. Families without children or with grown-up children may prefer a central location, in spite of the scarceness of space and its higher cost, because of its proximity and accessibility to cultural and recreational facilities, such as movies, theaters, and institutes of higher education.

On the other hand, families with many small children need more space, indoors as well as outdoors, and therefore may prefer suburban locations where land is cheaper and more space is available for children.

Dwellings should be designed according to the number of children in the family. When the number of persons living in the same house is large, duplication of services (bathroom) may become necessary. Also, the relation between sleeping quarters for the children and for the parents on one hand and between the former and the living room on the other should be planned so as to provide privacy for the parents and the opportunity for them to entertain guests without disturbing the children's sleep. Rooms for the children may follow two opposite criteria: a small but private room for every child, or a bigger room for more children. The decision in this dilemma depends on the family's attitudes about privacy and the parent's habits of child-rearing.

At the neighborhood level the needs of children require the provision of open spaces and play and sport areas in adequate proximity to the houses.

The age of children also influences preferences for various types of building. Several studies have shown that high-rise buildings are less suitable for families with small children because it is much more difficult for mothers to supervise their children when they are playing in the playground (for example, Hall 1966; Mackova 1968). The disadvantages of high-rise buildings are often accentuated in countries with hot climates, where private outdoor living space may be very desirable from the point of view of thermal comfort.

PATTERNS OF RECREATION

Differences in sociocultural background are reflected in preferences for different modes of recreation. The latter, in turn, call for different features of planning to insure a higher level of public utilization of recreational facilities, thus leading to more efficient expenditure of the resources allocated to them. Public recreation facilities, especially on the neighborhood scale, may also play an important role in initiating social contacts, particularly in reference to children (and their accompanying mothers) and teenagers.

Recreational facilities, especially public parks, are expensive items in an urban budget, both in terms of the direct expenses associated with their establishment and maintenance and the indirect expenses attributable to the loss of revenues from alternative land uses. It is important, therefore, that the usefulness of recreational facilities to the community be maximized. Benefits to a community in terms of a better sociocultural environment could be increased by organization of various activities that will enable the public to derive most from the physical facilities provided. In particular, the sociocultural variables of the population need to be examined in considering alternatives for facilities and their design details.

Some relevant sociocultural variables are:

- mass vs. individual recreation;
- frequency of various recreational activities;
- prevalent outdoor activities of young children with mothers;
- popular modes of recreation:
  —observing free public entertainment events
  —gathering in public parks
  —picnicking and outdoor camping
  —hikes in nature reserves
  —active participation in sports
  —observing mass sport events
  —swimming, boating, water-skiing, and other water sports
  —patronizing amusement parks
  —sitting in cafes and observing street activities.

Some planning features relevant to recreation are:

- total open area allocated for recreation;
- division of the total area between different planning scales: streets, neighborhoods, the city, metropolitan region;

- size of individual open areas at the different planning scales;
- facilities available for play and recreation;
- use of lawns, woodlands, ponds, etc. in public parks;
- details of design: pathways, location, and arrangement of benches, planting, opening of vistas, etc.;
- accessibility of the open spaces at the different planning scales: public transportation, roads and parking for private transportation;
- maintenance of the open spaces: admission charges, cleaning, repairing, etc.

SOCIOCULTURAL QUALITY AS A FACTOR IN URBAN ATTRACTIVENESS

Increases in the standard of living are generally associated with demands for higher quality and more variety in cultural life and education, and diversification of recreation and social opportunities. In fact, the availability of high quality cultural and educational institutions in a given city serves as a major factor in attracting people from higher socioeconomic levels (Berler 1970). This group, because of its professional training, has more freedom of choice and territorial mobility on the regional, national, and even international scales. Persons with professional training, because of their own demands and their desire to raise their children in an environment which will provide them with good opportunities in modern competitive society, are attracted to cities and neighborhoods offering such opportunities.

At the same time, the presence of a large proportion of high-level professionals in a city attracts more professionals and broadens the manpower base for the development of more advanced types of industries. Furthermore, professional persons contribute to the financial stability of a city through the proportionally higher taxes they pay. In this way, an investment of resources in the development of cultural and educational institutions may determine not only the image of city from a cultural viewpoint but also enhance its prospects for a higher level of economic development.

The sociocultural character of a city may also influence its attractiveness for another segment of the population which has a significant impact on its economy and stability, namely, the age group comprising the recent graduates of high schools and universities. The territorial mobility of this segment of the population is relatively high, because they are independent but not yet restricted by the burden of raising a family. The sociocultural aspects which attract these younger people may differ from those appealing to high-level, older professionals. Such facilities as movies, discotheques, youth clubs, and inexpensive dance places may be a major factor in attracting them to stay in or to move to a given city.

The role of sociocultural quality in urban attractiveness deserves special attention in the formation of policies at the national as well as at the municipal level for cities suffering from the flight of professional and other middle-class inhabitants. Saving tax money by cutting expenditures for cultural and educational facilities, in particular those that serve mainly a small, but better educated segment of the population, may ultimately prove to be a very short-sighted solution to the financial difficulties of the cities.

REFERENCES

Atkinson, G. A. House design and construction. In *Housing and urbanism, Inter-African Conference*, Nairobi: Scientific Council for Africa South of the Sahara, Pub. No. 47, 1959.

Berler, A. *New towns in Israel.* Jerusalem: The University Press, 1970.

Datta, K. L., and Lal, R. Social effects of dwellings distributions in multi-storeyed residential blocks. In *Proceedings, CIB Symposium*. Stockholm, 1968.

Givoni, B. *Man, climate and architecture*. London: Elsevier Publishing Co., 1969.

Gupta, T. N. Social and environmental basis for dwelling design. In *Proceedings, CIB Symposium*. Stockholm, 1968.

Hall, E. T. *The hidden dimension*. New York: Doubleday and Co., 1966.

Hole, W. V. Users needs and the design of houses—the current and potential contribution of sociological studies. In *Proceedings, CIB Symposium*. Stockholm, 1968.

Mackova, L. The results of sociological research carried out in the flats and the blocks of flats and their application in practice. In *Proceedings, CIB Symposium*. Stockholm, 1968.

Parry, D., and Stuber, F. High density living. *Connection*, 1969, **6** (1 & 2), 6–26.

Proceedings, CIB Commission W45, Symposium, Stockholm, October 1967. The social environment and its effect on the design of the dwelling and its immediate surroundings. Stockholm: The National Swedish Institute for Building Research, Report 5, 1968.

SEVENTEEN

Experience with Selection Tests in Various Countries

G. ARENDS

Philips has production units, or national organizations, in over sixty countries, where products of many different kinds, such as lamps, television sets, radios, medical systems, and data systems, are assembled by semiskilled workers. The factories abroad very often use the same tools, machinery, training systems, and quality standards as are used for the production of the same types of products in Holland.

The various semiskilled factory jobs throughout Philips are classified with regard to their level of difficulty by a "Functional Evaluation System," used for both blue-collar and white-collar workers. On analyzing the various elements contributing to the level of difficulty, we have found that one aspect dominates. Knowledge, experience, and ability needed for making decisions in various work situations, a description that sounds like a definition of "intelligence," has a correlation of 0.98 with the total estimated level of difficulty. Two other elements, "influencing people," and "dexterity in movement," have much less influence on the ranking of jobs according to their relative difficulty.

Thus, the work in our many factories is either of the same type or it can be equated in difficulty by our Functional Evaluation System. Comparisons of the output data, for example, quality and quantity weighted for the relative complexity of the product, could provide interesting indications of differences in productivity levels from country to country, if they exist. In actuality, the analysis of such indices as financial turnover, production in units, and efficiency per worker, has shown large unsystematic differences that permit no valid conclusions to be drawn. A more definitive answer to

G. Arends, Department of Industrial Psychology, N. V. Philips Gloeilampenfabrieken, Willemstraat 22A, Eindhoven, The Netherlands.

this complicated question would require a special investigation on output data from different countries in which the abilities of the workers and the levels of difficulty of the job are both held constant.

Here I shall focus on selection tests, the results obtained with these tests in different countries and cultures, and the relationship of the test results to work performance.

SELECTION TESTS

The Department of Industrial Psychology at Philips has a series of tests, partly manual performance tests, partly paper-and-pencil tests, which is available to all plants in the Philip's organization. If a department in a foreign national organization decides to make use of one or more of our tests, it must accept our conditions of uniformity in test layout and of training in test instruction and administration. It also has to furnish us with its test results, for which we calculate the norm scales, and its data on the work performances of the tested workers, so that we can calculate the predictive values of the tests. By this means we try to keep the selection procedure standardized and as completely under control as possible.

THE TESTS AND THEIR PREDICTIVE VALUES

Figure 1 shows the basic series of tests forming part of the "language-free" (that is, nonverbal) battery in use in about forty countries. Philips' psychologists in some countries have added tests of their own. For example, some verbal tests are used in Sweden, Germany, and South America.

Figure 2 shows the relationships between test scores and the performance of workers in the workshop in twelve different countries. The data are mostly for "concurrent" validations.[1] For each of the separate groups of workers who obtained a low, medium, or high test score, respectively, the numbers of those who were rated by their supervisors as fair to medium or good to very good workers are shown as percentages of their particular test group. Differences in work level are neglected, so that the rating "very good" within one test level group can refer to a janitor or to an assembly worker.

The results show that in countries with different cultures a standardized selection procedure consisting of almost the same type of tests can predict ability for comparable types of factory work.

NORMS

The test results collected in each national organization are sent to Eindhoven, where we calculate stanine norms for use in local selection

[1]In a concurrent validation, test scores are compared with ratings on jobs that employees hold at the time they are tested.

programs. At the same time, we make checks on how the tests are administered. Unfortunately, mean test results from different countries cannot be used for cross-cultural research. These scores depend so much on local situations, such as the general educational level of employees, the supply of labor at a particular moment, the image of the company as an employer, and the level of industrialization in an area, that only intranational comparisons are valid.

ANALYSIS OF THE TEST SERIES

Our analysis of the tests in different countries shows that almost all subtests are highly interrelated. All, except the marbles test, have correlation coefficients with scores on the total battery of between 0.50 and 0.70.

Factor analysis data for twenty-nine different populations are shown in Table 1. A principal-components factor analysis, with varimax rotation, was used in all cases. The results obtained with the Dutch data were designated "standard groupings." Names were assigned to the clusters from other nations to match the "standard groupings" as closely as possible.

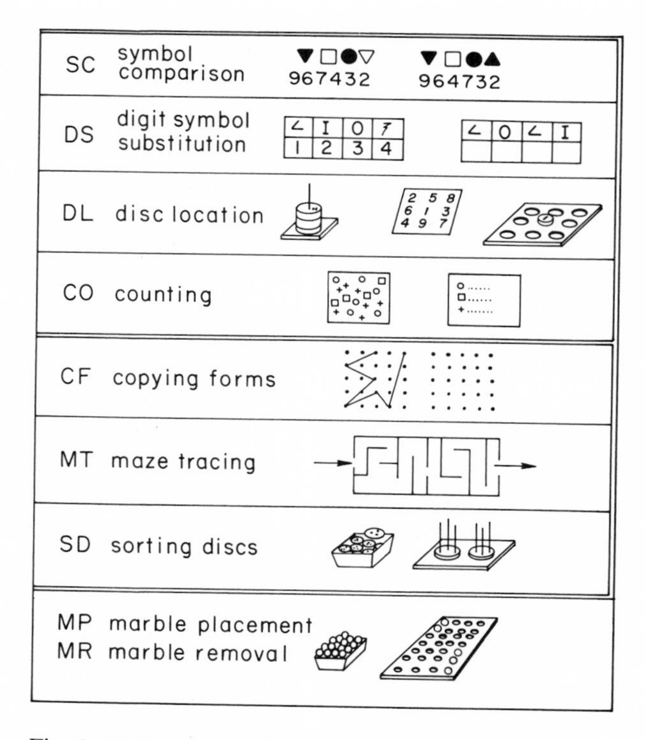

Fig. 1. Basic subtests forming part of the Philips language-free test battery.

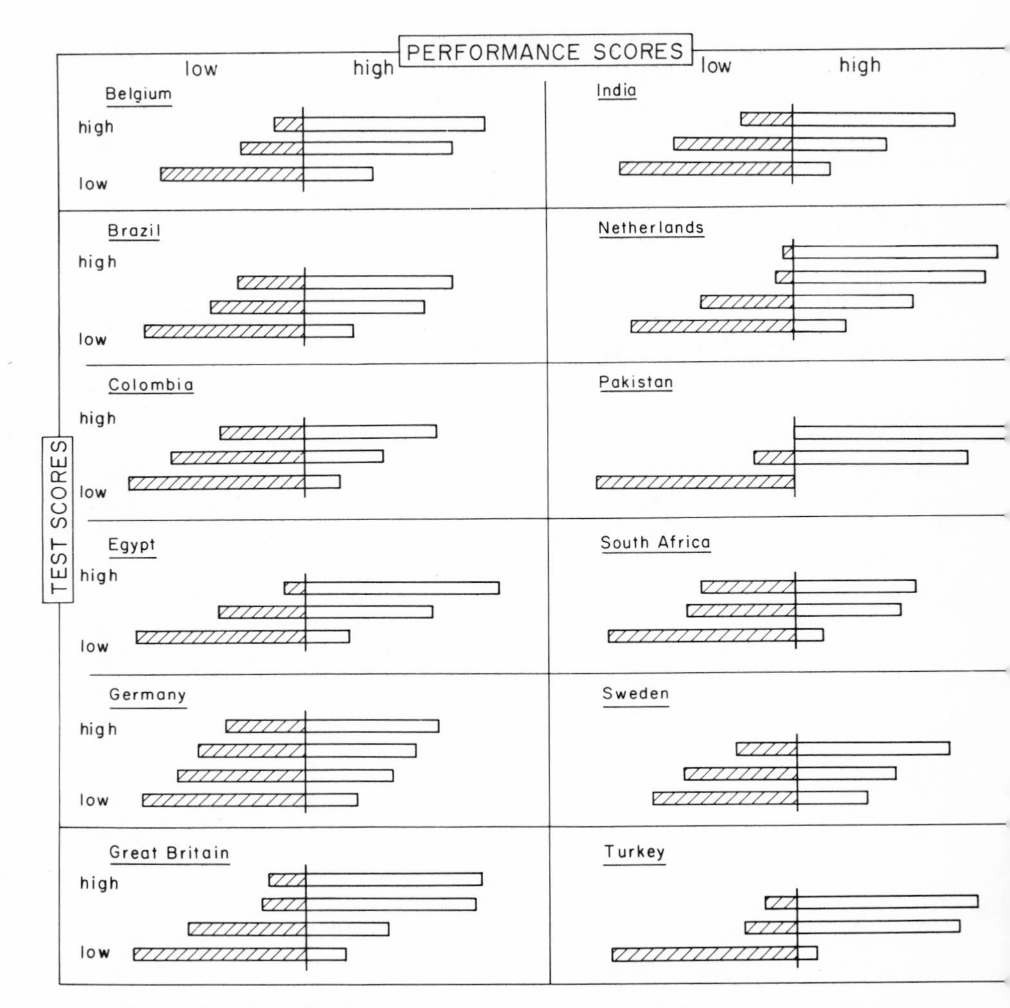

Fig. 2. Relationships between test scores and work performance in twelve countries.

Clusters that deviated markedly from the Dutch "standards" were given names that seemed to fit them best. Tests with loadings above 0.50 each fall into one or two of the following:

v	=	"verbal" cluster, i.e., tests in which a verbal component seems to be more prominent than other components
p	=	"perception" cluster
i	=	"insight" cluster
m	=	"arm movement" cluster
a,b,c	=	minor clusters, clusters not found in the Dutch data.

As is typical of most factor analyses, the names assigned to the clusters have a certain arbitrariness and should not be interpreted too literally. Blank spaces in Tables 1 and 2 indicate either that a test was not used in a particular situation or that the test has a loading of less than 0.50. Tests can, of course, occur in more than one cluster at a time.

The choice of one solution out of successive rotations in our factor analyses is not only subjective but also results in a loss of information about growth and changes in relationships between the variables, with progressive refinement of the analysis. To overcome this problem to some extent, the results of two successive rotations are given for the data from each country.

Inspection of the factor analysis data (Table 1) shows that in the majority of countries the contents of the nonverbal part of the original series consist of three clusters that may be described as:

perception: SC, DS, DL, CO
insight into shape and construction and eye-hand coordination: CF, MT, SD
repetitive arm movements: MP, MR

From this impressive consistency, we conclude that the solutions of our various test items appear to require the same type of mental processes in all countries. Moreover, these aspects of intelligence have proved to be good predictors of performance in factory work in all countries where they have been used.

EXPERIENCE, HYPOTHESES, SUGGESTIONS

From empirical data, from hearsay, and from contacts with people from different cultures we have pieced together some information that may be of interest to those who work with people of diverse cultural backgrounds.

TWO TRENDS IN THE FACTOR ANALYSES NEEDING FURTHER INVESTIGATION

In the data for countries outside Europe, the copying forms subtest and, to a lesser extent, the other tests for "insight into shapes and construction" often reveal the same characteristics as the perceptual cluster tests.

Hypothesis: Speed of accurate perception is generally less developed outside Western Europe.

Another trend becomes apparent when verbal test results are included in the factor analyses. This can only be done in the case of five South American countries (Table 2). The inclusion of verbal test results causes a

Table 1. Principal components extracted in the factor analyses of the data collected in various countries

	Netherlands	Germany	Germany	Sweden	Finland	England	Ireland	Portugal	Spain	Greece	Greece	Italy	Yugoslavia	Turkey	Iran
N	400	351	505	135	600	94	80	139	147	90	178	172	32	75	70
Verbal test items															
Qu[a]	v	v v	v v	v v											
Wk[a]	v	v v	v v	v v											
Wp[a]	v	v v	v v	v v											
Non-verbal test items															
SC	p	a p	p p	p p	p p	p p	pi p	ap p	p b	p p	p p	p p	pi pi	p a	p p
DS	p	p	p p	p p	a a	pa pa	p p	p p	p p	p p	p p	p p	p p	p p	p p
DL	p	p	p p	p p	p p	p p	p p	p p	pi p	p p	p p	pi p	p p	pi pi	p p
CO	p	i p	p a	i a	p b	a a	a a	a a	i a	i a	i i	p a	m a	p p	a a
CF	i	i i	i i	v i	i i	i i	p p	i i	p p	p pi	p p	pi p	pi pi	i i	p i
MT	i	i b			i i	p b	i i	i i							i b
SD	i	i i	i ia	i i	i i	i i	a b	a a	ia p	pi i	p pi	i i	i i	i i	b i
MP	m	m m	m m	m m			m m		m mb	m m	m m	m m	m m	m m	m m
MR	m	m m	m m	m m			m m		m m		m b	m m	m m	c c	m m

	Philippines	India	Taiwan	Egypt	Nigeria	South Africa	Argentina	Brazil	Peru	Uruguay	Chile	Colombia	Antilles
N	137	91	500	107	80	100	460	360	415	114	150	233	248
Non-verbal test items													
SC	p p	pi p	p p	p p	p p	p p	p p	p p	p p	p p	p p	p p	p p
DS	p pi	p p	p p	p p	p p	p	p p	p p	p p	p p	p p	p p	p p
DL	a p	i p	p p	p p	p p	mp p	p p	p p	p p	pi pi	p p	p p	p p
CO	a p	i	a a			p p	p a	p a	i a	i i	p a		p b
CF	p i	pi p	i i	pi pi	p p	i bi	pi pi	i i	pi i	i i	p p	p p	i i
MT		p b	p b	i b	i b	i i			b b	i i	i pi	i b	
SD	p i	i i	pi i	i i	i i	p b	i i	i i	i i	i i	i i	i i	i i
MP	a b	m mb		m m	m i	m m	m m	m m	m m	m. m	m m	m m	m m
MR		m m		m m	m m	m m	m m	m m	m m	m m	m m	m m	m m

[a]Qu = Questions; Wk = Word knowledge; Wp = Word problems.

Table 2. Principal factors extracted in the factor analyses when verbal subtest items are included in the test battery (*columns on the left*) and when they are not included (*columns on the right*)

	Argentina		Brazil		Peru		Uruguay		Chile		Argentina		Brazil		Peru		Uruguay		Chile	
N	460		360		415		114		150		460		360		415		114		150	
Verbal test items																				
Qu[a]	v	v	v	v	v	v	v	v	v	v										
Wk[a]	v	v	v	v	v	v	v	v	v	v										
Wp[a]	v	v	i	va	v	a			v	b										
Non-verbal test items																				
SC	v	v	v	p	v	v	v	v	v	p	p	p	p	p	p	p	p	p	p	p
DS	v	v	p	p	v	v	v	v	pv	p	p	p	p	p	p	p	p	p	p	p
DL	v	v	p	p	i	i	v	v	p	p	p	p	p	p	p	p	pi	pi	p	p
CO	i	p	p	a	i	i	i	i	v	p	p	a	p	a	i	a	i	i	p	a
CF	i	i	i	i	i	i	i	i	v	p	pi	pi	i	i	pi	i	i	i	p	p
MT					i	i	i	i	i	i					b	b	i	i	i	pi
SD	i	i	i	i	i	i	mi	i	p	b	i	i	i	i	i	i	i	i	i	i
MP	m	m	m	m	m	m	m	m	m	m	m	m	m	m	m	m	m	m	m	m
MR	m	m	m	m	m	m	m	m	m	m	m	m	m	m	m	m	m	m	m	m

[a]Qu = Questions; Wk = Word knowledge; Wp = Word problems.

complete change in outcome for the "perception" tests, which now show a fairly high relation to the verbal cluster. Contrast these results with those for European countries (Table 1).

Hypothesis: In addition to the above-mentioned lack of experience which results in slowness of accurate perception, there exists, at least in South American countries, a lack of experience in working with symbols, characters, and digits that influences test results. This lack of experience may be the result of differences in educational opportunities.

TEST INSTRUCTIONS OFTEN HAVE TO BE MODIFIED FOR DIFFERENT COUNTRIES

Some remarks from the leader of training courses for repair workers in Africa are instructive:

- Spatial and abstract topics are always especially hard to explain.
- I could not give written instructions. They did not understand them. Spoken instructions, sometimes repeated, are more successful.
- The best way to instruct is to visualize, for example, in my course for electronics I explained "resistance to current" by a drawing of a number of men trying to push through a narrow corridor at the same time.

- But not every drawing can be understood. It was completely impossible for me to explain your Block Design test to my candidates. They could not understand how a two-dimensional drawing could represent a three-dimensional pile of blocks, or how to count the invisible blocks. Demonstrations with matchboxes were of no help in explaining to them that there are also blocks inside a pyramid of blocks.

We found this last observation interesting. It is years since this test in which applicants are asked to "count the number of blocks needed to make the pile as drawn in the picture" was dropped from the test battery for semiskilled workers in Holland. The reason was that the frequency distribution of the results showed two significantly different peaks, indicating that we were undoubtedly sampling from two populations. The low-scoring group consisted mainly of female applicants, the high-scoring group mainly of male applicants.

Our hypothesis then was that as a result of culture patterns over past centuries in Western Europe, men were, by training and by experience in their jobs, more accustomed to working from drawings and thinking in three dimensions than women, whose production consisted mainly of two-dimensional embroidery and related tasks.

This lack of training in how to think three-dimensionally on the basis of a two-dimensional representation, that is, in abstract spatial reasoning, may also account for the difficulties we have had with the block test. We have now dropped this subtest in Holland and in Central Africa.

Another plant manager had the impression that his African workers engaged in inspection work had more difficulty in perceiving small production faults than might be expected. The remarkable number of mistakes we found in the digit symbol test and the symbol comparison test for such workers may be related to that observation.

All these findings refer to Africa, to which we have been able to pay a lot of attention recently. Comparable results will undoubtedly be found to exist in other parts of the world.

A final important observation is that experience with technical courses shows that less developed aspects of intelligence, such as perception and abstract reasoning, can be improved by intensive training.

MOTIVATION

In conclusion, I want to touch on another important factor that exerts a strong influence on the relationship between tests and behavior in industry. Several investigations in our Western European factories on the relationships between the capacities of workers and training results, reactions on attitude surveys, absenteeism, and labor turnover, have revealed clearly the importance of a man's motivation to work and to make use of the skills acquired by education or training.

Though the struggle for daily bread is the main driving force in many countries, changes in the rank-ordering of worker needs are certain to take place in the near future. These changes are being brought about by better education, more generally available information world-wide, increased societal agitation, and greater industrialization. Coping with this stormy evolution will require the multidisciplinary cooperation of economists, technologists, ergonomists, physicians, and social scientists, not only for education and training of the worker or a restructuring of his work and environment but also for changes in organizational structures, in types of leadership, and perhaps in societies themselves.

EIGHTEEN

Some Observations on the Teaching of Ergonomics

H. G. MAULE AND J. S. WEINER

The teaching of ergonomics, like its practice and research, is necessarily interdisciplinary and thus has problems that are different from those of teaching a subject based on a single branch of science. The problems arise from the need to mobilize the requisite knowledge and technology from a wide and traditionally unrelated range of disciplines and to bring them to bear on the issues with which ergonomics is concerned. The problems can be viewed as falling into three categories: organization, finance, availability of appropriate personnel.

Given sufficient finances, it should be possible to set up a single institution or department with the facilities, personnel, and equipment appropriate to each of the separate disciplines, anatomy and biomechanics, experimental and occupational psychology, applied physiology, and systems engineering, together with their supporting techniques. Although there are some notable examples of this approach, only a few technically more advanced countries have succeeded in establishing such organizations either for teaching or for research. In the United Kingdom, the Department of Ergonomics and Cybernetics of the University of Technology at Loughborough is an outstanding example. This one department covers almost the whole spectrum of teaching and research.

Another approach is to make use of the resources of a number of departments, schools, or colleges to build an effective curriculum in ergonomics or human factors. This is the solution we found at London, and it is a solution that has been used in a number of institutions in the United States, for

H. G. Maule, Director of Studies in Ergonomics; J. S. Weiner, Director of the Medical Research Council Environmental Physiology Research Unit, University of London.

example, Purdue University and North Carolina State University at Raleigh (see also Pew & Small 1973).

Still a third approach, and perhaps the one most widely adopted, is to teach ergonomics or human factors as an area of concentration or specialization in a conventional discipline, usually psychology or industrial engineering (Pew & Small 1973). While programs of this kind may be quite good as far as they go, they do not provide the broad background that is generally thought to constitute an ideal program in ergonomics or human factors engineering.

In this paper we describe our program as an example of the second approach to the teaching of ergonomics. This approach may well be used, even if on a less ambitious scale, in countries where for various reasons education cannot be carried on by a single department or institution. Finally, we venture some observations about how Western programs in ergonomics should be modified to adapt them to the needs of developing countries, in particular, India.

THE LONDON PROGRAM

In 1965 the University of London examined the feasibility of introducing a postgraduate degree course in ergonomics. Because of the "federal" structure of the University, which is made up of relatively independent colleges and schools, it proved difficult to develop a course on the basis of a single college. Nevertheless, within the University as a whole, there existed a number of departments in different colleges, each of which was teaching one or more of the basic subjects relevant to ergonomics. In the initial stages an analysis was made of the ingredients of an "ideal" course. Colleges and departments that appeared able to provide a part or parts of the course were invited to collaborate in the formation of an intercollegiate, interdisciplinary scheme. The essential structure of the course made possible by that collaboration consists of the following components:

- background, e.g., history of ergonomics, industrial organization;
- systems and control engineering;
- experimental psychology, e.g., perception, information theory, memory;
- biomechanics and anthropometry;
- work physiology and environmental physiology;
- occupational psychology.

In addition, close contact with industry provides case study material and project work.

Most of the classroom teaching and laboratory work is provided by departments already staffed and equipped so that they are able to take on the work without additional staff or equipment. Some research units of the Medical Research Council also collaborate in the teaching. Case study teaching is organized on a different basis, however. Collaboration is obtained from practitioners in industry, industrial or other research organizations, and from other individuals who are able to make a particular contribution.

CASE STUDY WORK

Case study work and project work form an indispensable part of the course. The case study is essentially a practical account of an ergonomic problem in which the teacher has been concerned. In most cases the teacher is able to describe the circumstances leading up to the identification of the problem, its solution, and an evaluation of that solution. Case studies are also provided by research workers and by those concerned with the day-to-day work of industry. They may be presented in the classroom, in the laboratory, or on the actual site of the problem.

PROJECT WORK

Each student undertakes a practical project under the supervision of a tutor. The study may be based in a laboratory or in industry. It may be a limited study of a narrow specific problem or a general investigation and measurement of, for example, environmental conditions in a work place.

STAFFING

For the first three years of the London course, there have been only two individuals whose full-time appointments are concerned with the course, namely, a director of studies and his secretary. The director of studies is responsible for the organization of the course. He arranges the teaching program, either through the departments and colleges or directly with individual teachers. He is in charge of all advertising, the initial stages in selection of students, and the initial discussions with students about their projects. He arranges the special seminars and numerous visits. The director of studies must be in close contact with what is going on in ergonomics throughout the United Kingdom and with grant awarding bodies. He has a special university advisory committee to guide and direct him.

EXTENSION TO OTHER COUNTRIES

The pattern we used at London has many features that may be practical in countries whose resources are insufficient to provide a full and self-contained ergonomics institute. This pattern is also appropriate for several reasons (for example, finances, administrative simplicity) in countries able to afford more ambitious and more expensive ergonomics institutes.

To establish a system such as the one in the University of London, two conditions must be fulfilled. First, there must be a university center with a requisite range of relevant departments. Second, there must be the possibility of close contact with industry.

The minimal range of disciplines required from the university are:

- engineering for systems analysis,
- statistics,
- experimental (human) psychology,
- anatomy or physical anthropometry for anthropometry and biomechanics, and
- human physiology.

In addition, contact with industry must be such as to provide a wide variety of case studies, visits, and project work. The importance of the industrial link cannot be overestimated, and the time devoted to it should be equivalent to 20 to 30 percent of the course time. To bring this into effect requires the appointment of a full-time director of studies with secretarial support.

Although the basic ingredients of ergonomics do not differ greatly from one country to another, there will naturally be differences in emphasis in different countries as regards the problems that most need ergonomic attention. This in turn influences some aspects of the teaching course, particularly the case histories, seminars, and project work. Differences in emphasis are evident even between such Western countries as the United Kingdom and the United States. Ergonomics in the former tends to be somewhat more physiologically oriented than does human factors engineering in the latter (Chapanis 1972). Variations in needs are even more pronounced when one considers developing countries. The observations that follow are based on the experiences of one of us (Maule) during an eighteen-month period of work in India.

CLIMATIC VARIATIONS

India, and probably other countries of Southeast Asia, South America, and Africa, needs to take special account of the high temperatures and high humidities that characterize both outdoor and indoor work. There is a rich scientific literature on the special physiological problems of working in tropical and subtropical countries, for example, Christensen 1964, and Hansson 1968, from the Indian Textile Industry. Even so we would suggest that there is still a lot of work to be done in this area, and these topics require much more attention in training courses in India than they should receive in Western countries.

THE BALANCE BETWEEN MANUAL WORK AND MACHINE WORK

In many developing countries, a very large amount of work is still done manually. This is particularly noticeable in the building industry, in civil engineering, such as road-making, in agriculture (Fig. 1), and in some forms of transport. Because of the large population and the serious problem

Fig. 1. Planting rice in the traditional way.

of employment, the diversion of resources into capital-intensive industries will necessarily be a slow process. Thus the role of ergonomics in India should continue to emphasize the "man-work" interface of the manual worker (see Fig. 2) more than the "man-machine" interface of the machine operator.

There has been some very relevant work carried out under Professor Lundgren in Stockholm on quite simple manual tasks. For example, he has demonstrated the loss of efficiency that may occur through failing to maintain an ordinary hand-saw in good, sharp condition. Also Hansen's studies of the ergonomics of the hand-operated wheelbarrow has clear application in a country where manual work dominates the building industry.

All over India men and women are lifting and carrying loads. The ergonomics of the tower crane, where it is used, is of course important, but the technique of manual lifting is of far more widespread concern in the daily work of hundreds of thousands, perhaps millions, of people (see Figs. 3 and 4). What do we know about the technique of lifting? What is the optimum size and weight of load? What should be the size of the receptacle into which concrete, sand, or grain is placed and then carried on the head? These are some questions that Indian ergonomics could solve with great benefit to the country.

HARD PHYSICAL MANUAL WORK

With the relatively small amount of mechanical power available in the developing countries, much of the manual work is of a very arduous nature. Examples of this include laying heavy electrical cables, transporting heavy loads by hand push-carts, hand or bicycle rickshaws, cane sugar cutting, and road-making (see Figs. 5 and 6). Basic information about the energy expenditure of such work might give surprising results and could lead the way to improved work methods. There is still a need for more information of this kind and the optimum deployment of manpower to perform it.

HOURS OF WORK

In 1895 Sir William Mather, speaking about the long hours of work in industry, said, "Of this I am assured that the most economical production is obtained by employing men only so long as they are at their best—when this is passed there is no true economy in their continued work." There are, we believe, still many instances in the developing countries of hours of work being excessive. Ganguly (personal communication) reports that over 20 percent of the average shift worker's time is spent in unofficial rest away from his work place when the official hours of work are excessive. In any

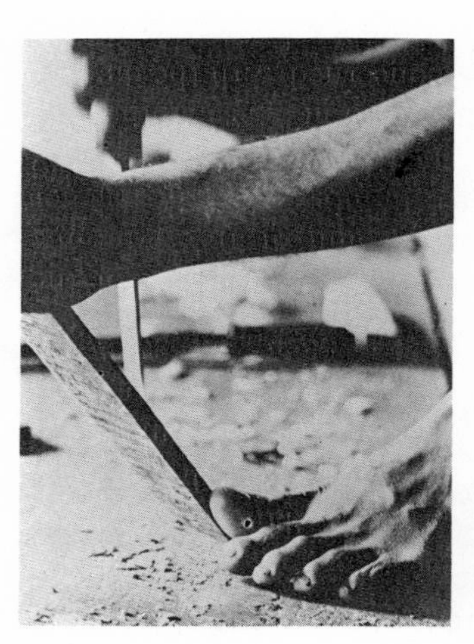

Fig. 2. Age-old techniques are still used in India. This pattern-maker holds the piece of wood he is working on with the aid of his foot. Safer methods are possible. (Photo courtesy of the International Labour Office, Geneva.)

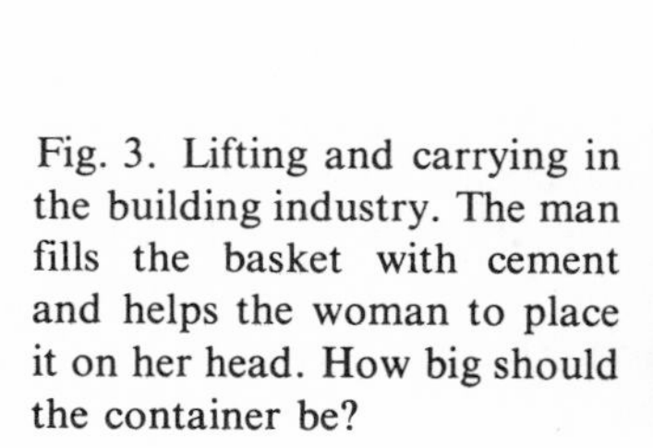

Fig. 3. Lifting and carrying in the building industry. The man fills the basket with cement and helps the woman to place it on her head. How big should the container be?

Fig. 4. These women are carrying liquid concrete to be used in the construction of a flat roof. Note the height of the head pad and the old rubber innertubes used as protective gloves.

Fig. 5. In road-making, women carry hot, tarred stones to the man who scatters them over the road surface.

Fig. 6. Tar being sprayed over stones in road-making. The man's posture and the position of the nozzles result in imprecise application and in a mist of tar being blown away. Could the spray gun be designed more appropriately?

case, this would appear to be a topic of much greater interest in the developing countries than in Western countries.

MOTIVATION

Although there may well be differences of opinion as to whether motivation and attitudes to work are legitimately a part of ergonomics, no one will dispute that motivation and attitudes are important elements in work effectiveness. We know very little, in the East or in the West, about the factors that make for high motivation and satisfactory attitudes toward work.

In 1964–65 the Hindustani Machine Tool Company completed a program for starting a watch factory in Bangalore. Indian technicians had been trained in Japan, and Japanese technicians were advising, instructing, and training at the Indian factory. Not only was the firm producing an excellent watch but it had also manufactured, under Japanese guidance, the first machine for making watches. Previously, all such machines had been obtained from Japan.

Thus there was a situation in which many Indian technical and managerial staff were thoroughly familiar with the same work in both countries.

These people described the completely different approach to work in India and Japan. Workers in the former country appeared to regard it as a necessary but on the whole unattractive way of occupying the day, whereas the latter approached their work as if they were dedicated to it. Although no figures are available, the attendance at, and attention to, work in the two factories was described as quite different.

This is, of course, a highly complicated problem, but these national and cultural differences may well be worth the same painstaking investigation that has been devoted to anthropometric variables. A better understanding of these variables may be more important to Indian industrial prosperity, and it may well be that such topics would be highly desirable additions to training programs for Indian ergonomists.

REFERENCES

Chapanis, A. Relevance of physiological and psychological criteria to man-machine systems: The present state of the art. *Ergonomics*, 1970, **13**, 337–46.

Christensen, E. H. *Manual work: studies on the application of physiology to working conditions in a sub-tropical country*. Geneva, Switzerland: International Labour Office, 1964.

Hansson, J. E. *Work physiology as a tool in ergonomics and production engineering*. Stockholm, Sweden: National Institute of Occupational Health, 1968.

Pew, R. W., and Small, A. M. (Eds.) *Directory of graduate programs in human factors* (second edition). Santa Monica, California: Human Factors Society, 1973.

Index

LIBRARY OF CONGRESS CATALOGING IN PUBLICATION DATA

Main entry under title:

Ethnic variables in human factors engineering.

Includes bibliographies and index.

1. Human engineering—Congresses. 2. Anthropometry—Congresses. I. Chapanis, Alphonse Robert Everysta, ed. II. Symposium on National and Cultural Variables in Human Factors Engineering, Oosterbeek, Netherlands, 1972.

TA166.E83 1975 620.8'2 74-24393

ISBN 0-8018-1668-8

THE JOHNS HOPKINS UNIVERSITY PRESS

This book was composed in Times Roman text and display type by
Jones Composition Company from a design by Susan Bishop.
It was printed on 60-lb. Warren 1854 regular paper and bound
in Joanna Arrestox cloth by Universal Lithographers, Inc.